a lucky
Prairie Boy

Douglas K. Brewster

Suite 300 - 990 Fort St
Victoria, BC, V8V 3K2
Canada

www.friesenpress.com

Copyright © 2021 by Douglas K. Brewster
First Edition — 2021

All rights reserved.

No part of this publication may be reproduced in any form, or by any means, electronic or mechanical, including photocopying, recording, or any information browsing, storage, or retrieval system, without permission in writing from FriesenPress.

ISBN
978-1-5255-8103-8 (Hardcover)
978-1-5255-8104-5 (Paperback)
978-1-5255-8105-2 (eBook)

1. BIOGRAPHY & AUTOBIOGRAPHY, PERSONAL MEMOIRS

Distributed to the trade by The Ingram Book Company

CONTENTS

Preface i

PART ONE

Opening 1

My Beginning 9

Jumping Ahead with Stories 21

Gibbs Farm Sale and Ontario Move 51

Growing Up in Earl Grey, Saskatchewan 59

University and Bachelor of Commerce 93

Becoming a Professional Accountant 123

PART TWO

Business Ventures 145

The Slow Down and Balzac Farming 179

Moving Back to Saskatchewan 191

Life's Reality 219

Lucky Me 239

Beginning Again 251

New Ventures and Family 263

Present Time 281

Acknowledgements 286

Dedicated to

My wife, Donna,
and our children,
Bettyanne, Stephen, Cathy

PREFACE

MY LIFE WAS SHAPED BY THE EARLY, raw environment of rural North America, not inhabited until 1886, when the settlers started to develop homesteads. Alone, gliding across the frozen pond on the beautiful, glassy ice—seeing its bottom clearly was like being in heaven for me, as were my two-mile trap-line trips, with paths through the snow between our prairie bushes in my trusty moccasins and a 22 gauge rifle at my side. These surroundings, with the harsh prairie weather conditions and family ties, were my world and left imprints on my soul that I will have forever.

I grew up surrounded by my dad's American heritage and the pride the Canadian members of his family had in their past and significance south of the border. My dad's Uncle Callie, a Doctor of Medicine in Brit Iowa, was a caring example of thoughtfulness to us and our Brewster relatives south of the border, helping to instill our proud connection with the United States of America. In looking back, I believe this pride was fueled by the knowledge and writings of our rich heritage, and the part our Brewster family played in the first settlement on Plymouth Rock. Genealogy books, family Bibles, and pilgrim documents all tell the story of

William Brewster and his group wanting to set up a society in the new world across the ocean. The want for a much freer society with individual rights came from his unhappiness with the English system of government and its controls.

I am so thankful for the efforts of our past Brewster relatives in researching and keeping our story alive. Many of them were members of the pilgrim society prompting the heritage and goals of the original founders. I took a keen interest in my American connection at a young age and have continued this interest through my adult life. In retrospect, I wonder if my enthusiasm for an active life and trying new things—sometimes at the detriment of family and friends—came from this Brewster unrest and wanting more.

I am the 13th generation Brewster to live in the new world (North America):

1. Douglas Kenneth (b. January 20, 1943) Regina, SK, Canada
2. Kenneth Willis (b. January 9, 1916) Earl Grey, SK, Canada
3. Maurice Richard (b. Febuary19, 1883) Ada, KS, United States of America
4. Irving Lyman (b. June 16, 1857) Blenheim, NY, United States of American
5. Lyman Horac (b. February 22, 1830) Ottawa, KS, United States of America
6. John Otis (b. April 1790) Blenheim, NY, United States of America
7. John (b. April 10, 1767) Blenheim, NY, United States of America

8. Daniel (b. April 12, 1731) Salisbury, NY, United States of America
9. John (b. July 18, 1695) Preston, NY, United States of America
10. Daniel (b. March 1, 1667) Preston, NY, United States of America
11. Benjamin (b. November 17, 1633) Duxbury, MA, United States of America
12. Jonathon (b. August 12, 1593) Scrooby, North Nottinghamshire United Kingdom
13. William (b. 1568) Scrooby, North Nottinghamshire United Kingdom

Although my family has a rich heritage south of the border, I also had a strong English influence from my mother's family (Woodcock-Carey)—a truly patriotic Canadian family.

Grandma Lille Woodcock (nee Carey) came from Fulham, England, a factory town, in early 1909. Grandma's family came to Kingston, Ontario, later settling in Peterborough. George, Grandma's father, served in World War I with the 109th Battalion CEF. Along with most of the family, he worked in Ontario factories, such as General Electric, Cockshutt Plow Company, Massey-Harris Company. Grandpa Woodcock (Edgar) was a descendant of Birmingham, England, eventually settling in Peterborough, Ontario. Together they had five children: Grace, Gwen, Hazel (my mom), Bert, and Edgar Jr. The family enjoyed sports with sons Bert and Edgar, and son-in-law Bill enlisting in the Second World War. Aunt Grace, the oldest child, become a missionary, working with the leper colonies in Bolivia. I marvelled greatly at her commitment to the teachings of God.

Not until much later in my life did I understand the extent that these two major family influences had on my life, giving me the courage, independence, grounding, and respect for others to be a good person, which I have always tried to be, and hope to have to succeeded...most of the time. I hope my life's story will demonstrate my enthusiasm and hope for making Canada a better place.

PART ONE

OPENING

THROBBING PRESSURE ROUSES ME. A SHARP PAIN of an involuntary swallow jolts me awake. Slowly, I begin to realize that my throat is lined with descending tubes. I calculate that I'm alive. I hear Donna's voice, so I know it is true. I'm living. I'm breathing in a world I know. The facts tumble in, one chasing the other. My heart is still beating. I hear its steady surge and swish. I did not die from this last heart attack.

"Donna?" I summon my strength to call my wife's name. My throat stings with pain so sharp I gasp. Dimly, I sense my mouth being pried open. I taste dry tubes and a smooth tube. I seek to orient myself. I know how to do this. I have done it before.

Tickling cables fade as, lying on my side, I gag inside Dad's grain bin. I squeeze and snake inside on my belly, as instructed. "Crawl on in there," says Dad. "Spread the grain evenly as it pours in. Evenly, Doug!" Here I am, a skinny young boy in Saskatchewan, small enough to crawl into our grain bin—the coffin-sized one with the arched roof. Dad urges me into the tight opening even as sharp bits of grain bite into the heels of my hands, even as, breathing congested, I obey. I am shaking. Into the darkness, I slither. First to one side of the grain storage bin and next, still trembling, to the other. I want out.

"Now?" Dad calls out as my small heart pounds. "Is it full yet?" Dust is up my nose. I can't breathe! Will I suffocate in dried yellow? Be snuffed out in a grain deluge? Buried alive? My heart drums. My hands sweat and my fingers swell. Extended, my arms become Apollo's wings, fanning the grain evenly into each of the four corners of the grain bin. "Swing your arms, Doug. Sweep the grain into every corner. No empty spaces!" My body stiffens as I gasp to obey my father in terror of suffocation.

—

Inside my cramped enclosure, dim lights and hoses seem to restrain me. Scratchy grain fills my mouth and nose. But someone else is moving inside this tight space. How could this be? Dad says I'm the only one small enough to crawl in here!

Discharged from the grain box, I am out. Fine dusts of grain clog my nose and needle at my throat. At last, I'm alive. I am alive! I stand up and I breathe in. It hurts to swallow. I take a deeper breath. I see a man wearing a pale blue jacket—not farm attire. "I'm your evening attendant," he says with a smile. "Your nurse will be here shortly."

Strips of soft cotton wrap secure the plastic tubing so I cannot yank them out in my sleep. There is a sudden thud. Oh! What fell? In a flush of panic, I monitor my whirring machine and note a new pulsation beating hard against the cable running up my nose and down my inflamed throat. Am I back in Dad's grain storage shed? *Get me out! Get me out, get me out!* I think but cannot say. Voiceless terror radiates stark energy from my head to my toes. Once again, I'm standing close to the unpainted wooden structure where Dad stores the oats he harvests, with the neighbour's thrashing machine, from the sheaves stoked in our field. The big stationary thrasher ran off a long flat belt connected to our Massey-Harris 44 tractor and made quick work of removing the oat grain from its stem. The straw was blown out of a long spout and into a large pile where us kids would later have great fun running and jumping off it. The oat seed flowed out of a pipe from the left side of the machine, going directly into our bin.

I know the year. It is 1952. Skinny as a rake, Dad deposits me into the sixteen-inch arched roof aperture once more. The jagged teeth of its corrugated mouth sample my ribs. I crawl on my belly, and again I wave my outstretched arms in a semi-circle, spreading the grain evenly to every corner of the bin. "No more!" I shout to Dad, without losing my life. Still, the sounds of Rice Krispies pouring out of a giant box begin. The grain auger rushes in the granules that soon rise to encircle my neck. A human tool, flat on my stomach, I yell to Dad, "Full! Full, Dad! Dad! Stop now! Stop!"

—

Machines tick beside my hospital bed. I am flat on my back, both arms trussed to rails. The motors around me, clicking and ticking, regulate the cables to which I am bound. I cannot speak. I'm all alone. The pain tastes salty. I am in panic mode.

—

This is not the grain bin. I'm next to Dad's Massey-Harris 21 self-propelled combine. It is plugged again, ground to a halt out in the north field. "Crawl in and yank out the blockage, Doug," Dad instructs. "What is it this time?" he calls after me. "Long weeds? Old roots?" Dad waits for me to crawl between the giant two-foot-wide blades. Each mud-stained cutting edge is ten inches high. Jagged rocks have already scratched silver signatures across their metal faces. As sharp blade edges taste my shirtsleeves, I enter my second place of terror.

"C'mon Doug. Move along, boy! Just a little further. Hurry and hustle on out." As I release the ropey brown blockage, two giant blades slip the tiniest bit downward. I cannot stop shaking.

—

"Better to lie still Mr. Brewster," said Nurse Maureen. "Breathe slowly now."

—

A little voice whispers to me as I tuck into my farm breakfast. With each swallow of milk, I rinse away prickly layers of grain dust, my throat red raw. I glance up at Mom and Dad. My little sister, Donna, smiles shyly. It seems like they do not hear anything, but I sure do. "Eat, Doug. Eat," the voice whispers. "Save your life. Grow too big to fit into the grain bin." I look up again. No one else seems to hear it. I dip into a second serving of steaming hot rolled oats topped with brown sugar.

—

My swollen hand slides along a steel sidebar of a hospital bed. Fingers fumbling, I pull up the thin ICU sheets to warm myself, if only for a few moments before I must get up for chores. Slowly, I turn my heavy head toward my window, where sparkling frost coats the sill and the icy edges of my bedroom floor, while snow sits on the top shelf of my closet from the night's north wind. Dad had installed storm windows and shovelled snow around the house, but my west-facing exposed second-floor bedroom, with its old siding and newspaper insulation, gives little protection. On hot summer nights, the room is an oven and I long for winter and my warm blankets.

Two more minutes of idyllic stretching before my day begins. I know morning's fragile light; it is 6:00 a.m. With no dawdling under my two Hudson's Bay blankets, and with a practiced glance, I assess the degree of snow and chill I'll need to battle. Before I leap out of bed to gather clothes already bundled, I reach under my pillow and grab two thick, warmed socks and knitted chill-guards that I pull on before hopping onto the freezing floorboards. I throw open my door and dash to the boiling warmth of the living room furnace.

Well before I woke, Dad had already confronted the dense farm freeze with his usual swift resolve. The chopped wood piled into his unique design is ablaze in the furnace. The centre area around the woodstove is warm. An hour before, lips compressed, Dad stuffed old, folded newspaper between the wood chunks and slipped a wooden match from the Eddy dispenser, the scratched tin decorated with an Alberta rose and fixed to the wall. Dad struck and held a single flame against one tip of dry paper. I wonder, now, if it gave him a tiny pause of satisfaction—an instant of gratification—to see the effect of his loyal efforts, the response of heat against his chilled fingers. I hope so.

Our comfort was guaranteed by Dad and certainly not by the indifferent prairie elements. At full speed, when I sprint out of my bedroom to the wood stove, my daily contest with chill's authority leaves no doubt of the weather's dominion. I had no idea, then, that Dad took the brunt of it. He is already out in the barn. I dress quickly. The biting chill lasts for only the time it takes to clothe myself in front of the blazing warmth Dad ensured. Each item of clothing mutes the chill: heavy, full-body long johns; two plaid button-down shirts (one larger than the other); denim pants; and extra thick wool socks. In the early days, my sister Donna, two years younger than I, was permitted to sleep a little later, so Mom could focus on lighting the kitchen stove so the side water reservoir would have warm water for our personal and kitchen functions. The kitchen, like my bedroom, was very drafty, with snow even getting into the pantry. Butter, or usually white margarine (it included yellow colouring not yet mixed), milk, and bread were often frozen, needing to be warmed for breakfast before being placed on the table. I partake of sustenance only after the regular sprint down our snow-covered path to the freezing cold outdoors, the morning chores now done. No dilly-dallying.

Animals first. I have my routine down to a science, as does any kid hungry for breakfast. The horses and other main livestock are tied in their stalls in the centre section of our huge, unique barn, with the calves, pigs, and chickens penned separately on the south side, below the hayloft that stores our hay and straw. It is a five-minute jog from the house, but I have gotten it down to three minutes.

First, I grab two steel buckets stacked at the west barn door, and then I charge to the wooden corner bin to fill one pail with ground oats, the other with barley. Swinging these on one arm, I grab a fork of hay, which after setting down the pails, I toss with one calculated thrust. I hear grunts and guffaws, bovine snorts and stomps—sounds of appreciation. The animals, as hungry as me, have far bigger stomachs to fill. On school days, I take care of these tasks, which spells a bit of relief for Dad. At eleven years old, now, I am also strong enough to give him a hand on weekends and school holidays, feeding and watering the livestock, cleaning out their stalls, pailing grain into the feed troughs, and in the evenings, bedding the livestock. Every day requires three to four hours of effort, but by applying my energy-saving techniques and challenging myself to be more efficient, I begin to calculate the number of minutes required to complete each task. Efficiency becomes a pleasant game, a race of competency and smarts. I like to win. I like the number of minutes that drop off as my speed mounts. I am grateful for my physical growth: expansion that guarantees I'll never again shimmy into a grain bin. Today, I hasten to my reward—a hot breakfast of porridge and homemade bread and jam—five full minutes earlier than last Saturday. Due to the cold winter weather, it was necessary to spend a full day of grinding grain and feed being stored in the chop bin and barn loft, for future daily feeding. I usually help, as the tractor must be started and attached with a belt to the pulleys of the tractor and hammer mill. Grain and feed are hauled, and hand fed into the rotating hammers. Proper lining up of the machines, using the flat ten-inch belt, is important for the belt to stay attached and run the hammer mill, which is quite difficult and takes practice. We are lucky that Dad's grinder blows the finished product directly into the proper chop bin, and our hayloft saves us a lot of manual work and time.

The winters are long, with freezing temperatures (below forty degrees Fahrenheit) and usually a great deal of snow and wind; however, temporary warming breaks usually occur, letting us complete the necessary outside work and prepare for another severe, cold, and snowy stretch of winter, just surviving and keeping the livestock fed and watered. Blizzards occur in all sorts of conditions, and the colder the temperature, the riskier it is for loss of life and the greater the suffering for all the living creatures on the farm.

In the house or barn, I feel protected from the howling wind, blowing snow, and little visibility. I do not consider the risks, especially the isolated situation of no available help or aid. The sixteen inch, cut to length Poplar tree wood pile near the house, which Donna and I stacked, is a sizable quantity against the north wall in our back porch, to keep dry, not having to uncover underneath the snow drifts later. It keeps the farmhouse warm while the livestock provide their own heat in the barn. The severe storms last two or three days with total isolation and the eerie sound of the northwest wind howling with creaking poplar trees. Emergencies are not an option, and somehow, nothing happens for the worst.

A survey after the storm is exciting. We see the large snow drifts on the south side of our bushes out in the field. Because of the cold northwest winds, sometimes trees are buried to their tops, with us being able to walk over them on the hard snow. A lot of work is created with the shovelling of snow from doorways and making new paths through and around the drifts, such as the dugout livestock-watering ice hole and the outdoor biffy.

Those of us involved and experiencing these severe weather conditions have a clearer understanding of our natural surroundings and its relationship with our overall environment.

MY BEGINNING

IN 1943, THE TRAIN TO MY FIRST home, forty miles north of Regina, Saskatchewan, chugged north from Regina through the Qu'Appelle Valley, passing by Craven, McKillop's landing (Valeport), Silton, and Gibbs on the new rail line completed around 1910, skirting Bulyea to the Earl Grey train station—still my favourite geography in the world. Prior to the railway, people and freight were moved by lake barge, stagecoach, and numerous well-travelled paths moving between farms and villages. The rail and service came to Craven in 1886; however, it was unreliable until the Canadian Pacific Railway (CPR) acquired the line after 1900.

My birthday is January 20, 1943—two years after the Japanese invasion of Pearl Harbor in Hawaii, home of the most powerful naval command in the history of the world, helped with the opening of the Panama Canal in 1914, both of which I visited in later life to learn about their significance and marvel at their construction.

Dad waited with a team of horses to drive us four miles in my very first prairie ride: an open three by twelve-inch sleigh box, piled thick with blankets to cover Mom and me, steering us across snow covered fields through our neighbour's fenced boundary lines, through gate openings and yards–Dad's quick alternative to uncleared roads heavy with snow drifts. He followed the snow trail to my grandparent's house. (I learned a few years later not to stand or sit still, but to keep moving to keep myself warm.) When travelling with our horse-drawn sleigh in the intense cold of winter, Dad would have Donna and I run behind the sleigh to help generate heat, so we did not freeze our feet and ears.

In my childhood, I would often imagine travelling in a boxy conveyance across vast, unfenced landscape. Soon after, around 1947, I learned about trains. Dad told me that when he was my age, during the Great Depression years of the 1930s, hundreds of easterners, suddenly unemployed, crowded the trains. Pushing away from the meager offerings at their kitchen tables, hungry men left their portions for their families and rode the rails to find work as farm hands in Canada's west. In fact, as a teenager, Uncle Calvin peddled his bike to Regina to employ and bring home one of those men, who walked the forty-five dusty miles to Earl Grey to help with their harvest.

A Lucky Prairie Boy

That day in my infancy, Dad drove the horse team to the solid one-story farmhouse of my grandparents: the formidable Hazel and the gentleman Maurice. Two tall windows stood at strict attention on either side of the front door of their unpainted small, square house. It was my first home.

I wielded a great deal of power from my cradle. Imagine that a desperate, squalling, illiterate, non-verbal, ten-pound bunch of muscle and sinew—me—convinced my grandparents, American homesteaders, and my parents, Kenneth and Hazel Brewster, to scramble as fast as they could to buy another farm. I engineered that purchase. My unfathomable screams night after night hastened my father's purchase of a farm ten miles southwest of Earl Grey and four miles southeast of the hamlet of Gibbs. The family's exhaustive adjustment to me, a melancholy and disquieting new baby, intensified the up-until-now politely contained personality clashes between the four sleep-deprived adults. Urgency soon escalated the move; exhaustion is a powerful agent for change.

There is a faded photo of the two couples, my parents and grandparents, on moving day. Standing beside my subdued dad, his unsmiling parents, and her bulging suitcase, Mom cradles me in her arms with a beaming smile on her joyful face. It has occurred to me that perhaps some politely repressed adult distress was loudly discharged by and through me, to Mom's welcome relief. I suspect my grandmother's forceful drive for work played a part in Mom's discontent, along with the cultural differences and being very homesick.

My parents' new farm, a grain and milk operation of reasonable size for the times (three-quarters of a section, south half of 14-22-21 W2, NE 16-27-21 W2), boasted nine milk cows and eighteen livestock. I have no doubt that Dad's family rolled up their sleeves and sunk many hours of work to help establish his farm. That Dad is a hard worker and smart builder is no secret to his family. Perhaps Dad's ceaseless effort was his non-verbal way of expressing gratitude for a chance to do useful work. Or was Dad obsessed with compensating for his father's shocking 1929 losses? Compelled to restore financial balance, perhaps Dad could think of nothing else. Sudden loss can and does derail one's own goals in service of restoring family stability. Children's dreams can be shaped or sacrificed, for what is perceived as the urgent, greater good?

That summer, Dad purchased a Farmhall H International tractor with grating steel cleated wheels. "Dad, the wheels have no rubber! Even the animals bend their ears. No wonder they're high strung lately!" I said. The rocky ride on screeching steel wheels over the frozen stone ground testified to war-time minimalism.

"The factories are serving the Second World War effort now, son. Farm equipment comes last, even though western Canada feeds the troops," said my dad.

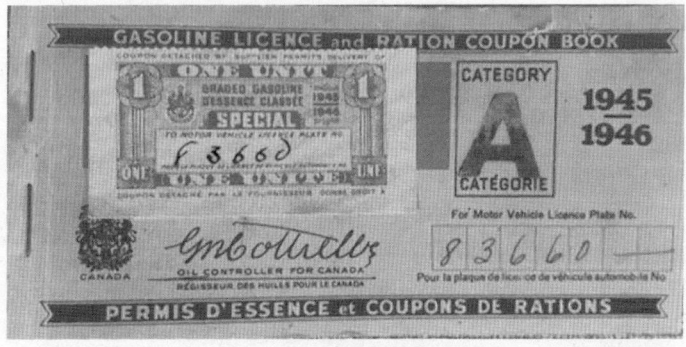

Ration Ticket

Later, rubber tires that held air were available and installed on our tractor.

After supper, Dad lit his kerosene lamp by removing the shade and lighting its mantel. He turned on the battery-run radio and spread out the weekly farm newspaper on the kitchen table to learn about what was going on in Canada and Europe. One evening, forgetting to reinstall the lamp shade, he leaned in too close as he went to turn up the volume. Pulled in by the disturbing rumors about exterminations, mass graves, and bombs dropped in England, Dad's hair caught fire in the lamp's flame. Yes, there was a terrible war raging a world away. Dad knew that, because of his farm produce going to Europe to feed the troops. With the establishment of the Canadian Government Wheat Board for the Western Prairie Farmer, gas rationing affected him and his neighbours, along with limited supply of farm equipment, farming was not easy during that time.

Born during the World War I, my dad's trauma began when he was twelve years old. One autumn morning, in October 1929, young Ken thought he heard an animal groaning in the barn, but the high-pitched cry was too close. Stumbling down the dark hallway, Ken found his mother, who was usually slicing homemade bread at the table, bent over its oilcloth, his father silent beside her. His arms were limp at his sides as he stared blankly ahead. Unnoticed, Ken listened.

"We've lost everything, Maurice? Is that what you're saying? The bank owns everything we have, everything we worked to build, everything? Please, Maurice, tell me this is a terrible mistake!" Ken overheard his mother say to his father.

Then Ken heard the comfort he banked on for the rest of his life. From his father's lips, he heard two words: "Nearly everything."

Nearly. That meant there was something left, an amount to increase. That single word spelled hope. It was settled. Ken would forever do his part to restore the family to stability.

So yes, Dad was a committed, nose-to-the-grindstone, frugal farmer, reluctant to throw caution to the wind and risk all for any banker's rosy-slick promises.

Dad and his family also suffered additional winds of misfortune, both real and symbolic, during the worst economic crises of the 1930s. How to describe the subsequent betrayals. How to comprehend Mother Nature's generosity on the one hand and her treachery on the other. What to do when windy weather reduces full grown crops to fine dried meal, with trillions of grasshoppers per square mile with their small jaws that chew every square inch of field to dehydrated dust. Dad watched his mother empty four full dust buckets every morning, filled to the brim with gray powder that sifted under the doorways and through the chinks of their walls every night. Even years later, as a married man and father of three, Dad's worn expense notebook testified to his early trauma. He set his to-the-penny accounting ledger and his short, knife-whittled pencil in the first drawer to the right of his worktable. We knew not to touch it.

—

I have no recollection of a horse-drawn buggy as the summer mode of our family's transport, or of my bicycle rides in an apple crate strapped down behind my mother to the Fiessel farm. However, horses still played an important role as a mode of transportation in rural Saskatchewan, especially in the winter and spring seasons. The family cutter and sled served a valued function in the winter. We needed it. Whether or not we owned a vehicle, we could not drive it on the impassible winter prairie roads. So we were grateful for our four-legged conveyor whose only gas fill-up is chewed oats.

We had neighbours that treasured their horses and spent hour upon hour of their precious time fussing over the animals: brushing their coats to brilliance; braiding their sleek, oiled tails; and feeding them like royalty. Brass insignia decorated their polished carriages, elegant conveyances piled with classic woven tartans and engraved brass-tipped reins. Such status symbols, refined but pointed references to wealth and position, suffered harsh reversal when shiny, painted mechanical tractors bullied their way to the topmost rungs of glory. While Dad had no time for braiding his horse's tail, he longed for the most modern steel farm equipment that would cut his workday in half. He enjoyed cleaning and servicing it with pride. Soon enough, the gleaming rubber-tired, self-powered machinery would assume its mighty place in history, and horsey status symbols, no matter how elegant, would topple to a designation of the past.

—

Mom, Donna and I were standing on the front porch steps, our mouths unhinged as Dad rolled up with our first farm vehicle: a Model A Ford. Before our wide eyes, he flipped open the trunk lid and invited us to try out the extra back rumble seat, which he later laid flat to carry large milk tins (twelve by thirty-two inches high) to the Gibbs train station for their trip to Regina. Little Donna was ecstatic at the roar of the engine and pivoted her small body to stand up inside our new vehicle.

"Hold her, Hazel," Dad said as Mom grabbed her intense toddler.

"My, our little Donna Grace takes after your mother, Ken," said Mom. "Talk about nimble n' grit!" she said with a laugh. "If our energetic lass had arrived at Gramma's house back in the day, instead of this city girl, Gramma Hazel would have met her match!" This first girl to the Canadian Brewster family was a happy occasion. No one anticipated the mischievous, strong-minded, capable girl and caring adult she would grow up to be. Donna was like her grandma, Hazel Brewster (nee Flint), whose mother's side of the family came from Minnesota, USA.

"Yes," Dad chuckled, "Our Donna is going to rule the roost! Of course, remember that my mother was an only child. There were no siblings to contest her authority!" said Dad.

"To be sure," Mom replied as my little sister wriggled in her lap.

During this special tour, Dad celebrated his grand purchase: the model with the first letter of the alphabet standing proud next to it, the A that reports excellence. He never once compared his shiny new prize to our black team of horses, though. He never diminished their animal value. Instead, he recalled to me their invaluable sleigh service in the winter of 1943, the year of my birth. "Once you and your mother were safely home, Doug, that spring, I placed the grain box that carried you home on top of our steel-wheeled wagon." It was used for hauling product in the summer, I learned. My dad placed value in multi-usage in all things.

Dad's inaugural drive around our farm wound around the milk barn with our solid stone well. One summer's day, Grandma Brewster told me of a similar stone well in Plymouth, Massachusetts, built on the land of Elder Brewster in 1621, a year after the passionate Puritan minister and his Mary stepped off the Mayflower. Grandma told me that by the time I got to see that stone spring, I'd be able to read the sign over it that reads, "Drink here and quench your thirst. From this spring, pilgrims drink first."

Dad circled the outside water tank, which runs non-stop to cool the full milk tins between deliveries to Gibbs. "It's going to be easier now," said Dad happily as he drove around the chicken and turkey buildings to tour his wooden granaries. With no method of refrigerating the milk and cream at the train station, our schedule was carefully coordinated to avoid costly spoilage.

After all, milk fetched $0.62 a gallon, three times the cost of gas at $0.19 a gallon.

Mom smiled at Dad as she climbed out of our newly purchased used car at our sturdy no-frills farmhouse. "Wait, Hazel," said Dad. "How about we zip over to show my parents?"

"You go, Ken," said Mom. "I'll get dinner ready. No point delaying the chores." She closed the car door and headed inside, her two munchkins in tow. I wanted to see Grandma, but I had to go where Mom went.

In 1946, when I was three, I decided to take my pretend milk tins to Gibbs. I headed north on the melting spring sleigh track. My frantic parents and neighbours searched the countryside for me. When I was finally found, my simple explanation was I was on the way to Gibbs to meet the milk train.

—

With another flash of my red scarf, wet snow splashed my face. There she went! Mary, George Brandt's much-loved youngest sister, was driving her magnificent one-horse cutter to school in Gibbs. She flew through our snow-covered yard and onto a trail winding through our glistening farm fields. What a Christmas card! Our neighbours, the wonderful Brandt family, lived across the road to the south and loved having their last child still with them. Deep snow and drifts caused by bitter northwest winter winds rendered the road impassable. Spring thaws and summer rains turned dirt roads to thick, suety, mud-rut trails. Originally flat, un-graveled, and with a two-foot ditch on each side for water run-off, it was not until the mid-1950s that most roads were raised three feet and given a little gravel. This was the start of "weather roads." Given this isolation, the Brandt's soon become family— quick to help us in a pinch, as we were for them. Whenever Dad hit a troubled patch, our neighbours ran over lickety-split. Even with blood family the next town over, a visit involved careful consideration of animal care, milk timing, the cost of gasoline, and road and weather conditions. The visits to Dad's family, six miles away, were infrequent, other than holidays such as Christmas and New Year's. To me, the word "family" was an elastic noun that stretched wide to include our neighbours.

—

Like my sister Donna in the early days, our black horse team were high strung. They had a reputation for being terrible kickers and hard to handle. To be trampled upon or bitten by the grinning yellow teeth of a thousand-pound horse is no laughing matter. Yes, horses are majestic beasts, but not all horses are well-behaved. They have strong emotions, just as we do. Well-aware of their threat, Dad found handling and harnessing the horse stressful. Tied in the barn, side by side, Dad never went between them, afraid of their kicking him with a sharp hoof. Dad told us never to get near a horse's back feet. He always led his mounts out of their stall and well into the barn alley before harnessing or watering them.

One day, our jovial neighbour, warm-hearted George Brandt, came over to visit. He slapped Dad on the back and bet him he could safely walk between the two horses tied in their stall to the manger. I stiffened. My eyes were wide as I watched Mr. Brandt do just that. What a risk-taker, always up for a new challenge. Dad shook his head in mock disapproval, a small smile forming. Once harnessed though, our proud animals performed with stamina and speed by pulling hay racks, sleighs, or wagons.

George Brandt loved horses. Dad respected them.

—

"Here you go, Doug," said the magnificent Bob Heckman, a tall, thin merchant who delivered our much-needed gas. A man of few words, he began by dropping several old tires from the back of his one-ton blue Chevy truck, with its six feet by eight feet box, onto our dusty designated spot. Then he rolled and tossed the four forty-five-gallon drums of gas onto the rubber shock absorbers, which eased their impact on hard ground. I admired his practiced skill as he rolled the heavy barrels of fluid and placed a four-inch rock under the bottom of each drum. He slipped the bungs crossways to keep moisture out, tipped his Eddy's Matches cap—the one with an extra wide visor—and tossed his soft tire rugs back onto the truck. With a smile and nod, he climbed into his truck cab. If Dad asked him to put it on his tab, Bob would assent with a slight nod as if to say, "No big deal," and then be on his way.

"Doug," said Dad, "without the kindness of that fine man, I'd be in deep trouble. He's saved my bacon more than once." Later, at supper, Dad again voiced his appreciation for Bob Heckman to Mom. "If not for having time to pay, we'd not survive, Hazel."

"Bob Heckman deserves the Stanley Cup of the west," said Mom, a devoted sports fan, along with all her Peterborough, Ontario family. "The Stanley Cup, Doug," Mom began, proud of her country's sports history, "was commissioned in 1892 as the Dominion Hockey Challenge Cup. The trophy itself is named after Lord Stanley of Preston, the Governor General of Canada. He donated it as an award to Canada's top-ranking amateur ice hockey club. So, yes, Bob Heckman is our Lord Stanley, to be sure," Mom concluded, passing a bowl of hot buttered green beans to me. Mom and Dad glanced at me when I added, "Mr. Heckman's our hero."

For his part, when Dad's money came in, Dad raced to Bob's store to pay off that good man, first and foremost. He would not wait to do it.

JUMPING AHEAD WITH STORIES

I AWAKEN HOURS LATER. A SENSE OF gratitude drizzles in, like melted sugar, along with the intravenous tube the nurse is adjusting. Transparent sacks hanging from an overhead rung are feeding me. I am still alive and survived another heart attack, thanks to Dr. Edwards. Here I am, now, a seventy-five-year-old retired farmer and businessman, husband of forty-three years to an outstanding woman, father of three wonderful adult children: Bettyanne, Stephen, and Cathy. I am a grandfather to Bettyanne's three children, Jessica, Callie, and Luke, and Stephen's two children, Colton and Camden. I view my dim room and pick up the repeated sounds of thrum and click, gurgle and tap, tap, tap. I monitor the wonderful care I'm given, thankful for the medical care I receive as a Canadian citizen.

A familiar touch of warmth, a gentle pressure, and a voice instantly bring me to full, joyful recognition. I am awake. I see her warm smile. I feel her love. Instant tears sting my eyes. Donna: my wife, my best friend, and the treasure of my life.

"In the morning," says Donna, "you'll be allowed a tiny taste of breakfast!" My mouth unavailable, she kisses my tube-free head and steps aside to reveal our beautiful granddaughter Jessica, who is now a sparkling Saskatchewan schoolteacher. Donna and I agree that all three of the grandchildren Bettyanne has given us are exquisitely polite, socially adept, and caring. When Bettyanne recounted the time young Luke stopped to change a flat tire of a distressed driver on the highway that week, we nodded. Of course, he would.

"Hi, Poppa!" Jessica says with a laugh. "Shhhh, don't tell anyone, but I've ordered a yummy breakfast for you to eat in the morning. Guess what our secret treat is, Poppa?" Jessica smiles, encircling my swollen hand with warm fingers.

"You still remember?" I ask, so happy to see my creative and energetic granddaughter, now a dedicated professional.

"My memory, Poppa, includes hot porridge and lots and lots of brown sugar. We would always wake up early when visiting you and Nana and get our sweet treat before our parents woke up. I loved the tiny grin on your face as you enjoyed watching us eat the sugary cooked oats. When Mom and Dad woke up, we all pretended that nothing happened. It was our secret!"

Still holding my hand, Jessica welcomes my second visitor. I notice my son is wearing the pioneer shirt I gave him for Christmas when I told him our forebears, William and Mary, arrived in Plymouth, Massachusetts in 1620. His hat is in his hand.

"Dad?" he whispers.

"Stephen," I reply, deeply comforted. Looking into my son's steady gaze, holding my granddaughter's hand, I relax. Everything feels better. I tell Stephen and Jessica what I told Donna. "I'm so happy to be alive!" Stephen understands my meaning, however inarticulate I sound.

As Jessica slips out to see her new husband Danny, a nice fellow who just completed his auto mechanics diploma. My son, a handsome, good man, settles himself into the deep armchair close to my bed. He opens his mother's notebook as I doze off once again. I'm glad he is with me.

The following morning, I am excited to learn that before my tiny taste of hot, brown-sugared oatmeal, there'll be the removal of two of four tubes in my mouth. I will be able to speak better and properly pronounce words. To see my son smiling at my bedside and to know Donna is on her way over with our daughters is great happiness. Tomorrow, my bed will be raised two notches. I know I can handle the increased pressure and pain.

"But Stephen," I respond to my son's question about his grandpa, Kenneth Willis. "Your grandfather, my dad, lived through a severe depression in 1929. You are a banker, so you know what that collapse meant to every living soul in the west at that time. There was no selling of cattle for a while. There was eating them out of necessity though. Why, I recall asking my dad if I could see a movie at the Earl Grey Hall one weekend and watching his serious face fall to calculate, with his trusty pencil and notebook, the cost of the movie ticket and gas to get me there and back. At that moment, I learned not to ask.

"Dad sunk every penny he earned into farm products to grow his operation. Our town hall most Saturday nights would get movie reels that would run on a projector behind and over the crowd, showing the image on a white screen in the hall front. The town hall was used for various functions, so movie goers sat on hard stacking chairs and would occasionally be interrupted by a film break that the local operator would have to repair. And after the movie, everyone stood and sang 'God Save Our Queen.

"I know Grandpa worked so hard, Dad. I never once saw him throw his head back and laugh, with his mouth wide open and his teeth showing," said Stephen.

"He'd smile at a good joke, but usually his lips were compressed, like the laughter needed harnessing. In looking back, I sense that at the outset of the war, when I was born, Dad was still so affected by the financial devastation that I honestly don't think World War II made a deeper dent."

—

"Hi, Poppa, it's me again," said Jessica. "Remember you and Nana making us real ice cream one night, bundling us up warmly and taking us outside into the dark to chip ice by moonlight? We were in the north field marsh. On our slow walk home with a full pail you told us about making the best vanilla ice cream years earlier with your friend Ron Fiessel and his parents, Chris and Elsie. It was the freshest, most delicious I ever tasted, Poppa. I'll bring a scoop my next visit, okay?"

"Can't wait, Jessica!" I reply.

—

"Honestly, Stephen, my life in the 1950s, with little communication outside the farm, meant no real idea about government—good, bad, or indifferent. Our family's existence was less about politics and more about improving life on our farm and helping our closest neighbours.

—

"You know, Cathy, I'm framing a scene in my mind right now. I am in my Grade 8 classroom, looking out the window. I see Dad driving his brand-new green tractor across the bright yellow prairie wheat field. He is sitting tall and driving the gleaming machine, it seems to me, with considerable pride. It is a grand moment for me. I am proud of Dad's new John Deere 70 diesel-engine tractor.

"Tell me, Dad," Cathy began, coffee in hand. "At that precise moment in time, do you think your dad would have liked to trade places with you? Maybe him in the classroom seat and you on the tractor?"

"No," I said, followed by a louder reply. "Absolutely not! I see now that while Dad respected classroom learning, he'd never have traded being his own boss in his wide-open, fresh-air classroom, not for a single second." Two images flash through my mind: one of Dad, rag in his left hand, polishing the machine he kept so clean; the other of his gnarled right hand adjusting, as he constantly did, the hand clutch and pulley of his dream tractor.

"I wasn't crazy about school either, Cathy. It still seems a bit of a shame that young kids are corralled so early in life. Still, I did not hate it. I met nice kids, which sweetened the deal. I did what was expected, and I learned to tolerate the boredom. But still, Cathy, who knows, maybe that tedium sparked my imagination."

"Mom says you were an entrepreneur ahead of your time, Dad, and I agree."

"Did she say that, Cathy?" I smiled my pleasure.

In the silence that followed, I found myself considering my Grandpa Maurice's history, seeking one plausible reason he would have bartered a cash-advance crop yield deal in 1928. Why did he haul off and do that?

"Cathy?"

"Yes, Dad, I'm still here. What are you thinking?"

"I'm thinking your great grandpa Maurice was a hero in his own right. Imagine a farmer with only one eye, Cathy. In 1901, when Grandpa Maurice was nine years old, he lost an eye on a single jagged, uncut edge while closing a heavy wire gate all by himself. That must have been awful pain! Imagine how long it must have been until he was heard screaming out in that yard. Was there medical care nearby? I hope he was given a shot of whiskey for the pain!"

"I can't imagine, Dad."

"It was 1901 on a farm in Kansas, USA. All we know is that he survived the grim wound," I continued. "Three years earlier, when he was six years old, he lost his mom, a schoolteacher, to sudden death. Born in Ada, Kansas, he was raised by his maternal grandmother James, with his older brother and sister…no history about his father. I do know that little Maurice worked with the energy of a grown man, much like his son was destined to do in the future.

"And, I might add, you too, Dad," commented Cathy.

"Maurice, the lad with an eye patch, had no idea on his fourteenth birthday that he'd soon move to the promised land, the ninth province of Canada, established on September 1905. He had no idea that Alberta joined Confederation that same month, or that by a Royal Warrant of King Edward VI, Saskatchewan, a province he would soon homestead, would be granted a Shield of Arms. Few youths know their future, and even fewer understand their family's history.

"One of Maurice's employers in Kansas claimed he didn't need to waste money on one of those fancy, new-fangled corn-picking machines, because that fourteen-year-old kid with one eye worked harder than any contraption could."

"That comment must have warmed his heart, eh Dad?" said Cathy.

"Oh, I'm sure it did, Cathy. A little approval goes a long way for a kid. And there is no question about the quality of care Gramma James gave her beloved daughter's child. Maurice never ever referred to himself as an orphan. When asked about his mother, he would point to his Gramma James, a woman who dearly loved her grandchildren."

"She reminds me of Mom's mother (Nana) Dad. She loved us hugely!"

"I was a regular at her finely laid table when dating your Mom, Cathy. Man could your Nana cook! Her roast beef and Yorkshire pudding, followed by her warm apple pie with vanilla ice-cream, was heaven."

"I know, Dad. Mom says you never turned down an invite from Nana."

"Of course not. I'd not wish to offend such a great lady," I said with a smile.

"Absolutely," Cathy said, smiling in return.

—

"So, I'm thinking," I said later, picking up an earlier thread of our bedside chat, "A farmer with a serious handicap might dream of a little extra support, right? An investment in hope."

"I agree, Dad. I mean, what if Great Grandpa Maurice lost the other eye? I am sure he was extra careful around farm fences and took every precaution to avoid a repeat or another accident. But what if? Then what? I mean, Maurice's margining deal was well before health and unemployment insurance were established. Maybe he was taking care of business as his gramma James taught him to do."

It must have been tough for Maurice and his adolescent son, Ken, to see the prosperity of the promised land turn to dust. The ninety percent loss of three hard-earned farm crops; fifteen years of slog. I heard tell that under a full moon, in the dead of winter, the sounds of Maurice chopping firewood sounded far into the night.

—

That afternoon, Stephen arrived, his wife, Amanda, at his side.

From a second manila envelope, I noticed my son withdraw a faded sepia coloured photograph of his great grandmother, Hazel Willis Flint, a talented hat designer. The reverse side is stamped with the photographer's business name and the year 1914. Her direct gaze is beautiful.

"It seems my great grandfather put the one eye he had to excellent use," Stephen said, "an excellent capacity of Brewster men." Offering Amanda the better chair beside him, he resumed an earlier discussion, one he was eager to hear now that his father's attention was guaranteed. No meetings, no trips, no distractions.

"Great Gramma Hazel responded to the love letter Maurice sent to her, and the two were married on Valentine's Day, 1915. She gave up her millenary work in sunny Berkley California to become a farmer's wife in Earl Grey, Saskatchewan, Canada. That must have been some letter-writing skill."

"Maybe she knew a good man when she saw him," said Amanda, smiling at Stephen.

"Yes, Maurice must have been convincing, son."

"And caring. He treated her to a piano, I hear. Quite a nice gift!"

"Yes, Gramma Hazel loved singing and playing the piano, and in 1916, for their second anniversary, Maurice had the instrument shipped to Gibbs. Although Dad was not yet born, the story goes that shortly after the gift arrived by train and was hauled home by a team of horses in a wagon box, suspicious neighbours reported to the authorities that guns had been imported from the United States. Soon thereafter, an officer from the North West Mounted Police rode into their yard, dismounted, inspected the piano crate, entered the house, examined the keyboard, sat down, and played for a while. Then he got up and rode away on his trusty steed."

"And your dad was born that same year, right, Dad?

"Yes, and named Kenneth Willis, after his mother's father, Willis C. Flint. Seven years later, in 1923, Ken's brother, Calvin Maurice was born. About the name Calvin, I discovered that just as William and Mary Brewster, our Puritan forbears admired the Frenchman, John Calvin. So too did Hazel. His religious books were found in her home library.

"A Puritan reformer, Calvin believed rewards are earned both here and in the afterlife. Your wealth signified that God was rewarding your goodness. If you were poor, Calvin believed that was punishment for your personal failings. If misfortune befell you, he believed you need only to look in the mirror for an explanation."

"Quite harsh, no?" said Amanda. "A dust-bowl-drought for a little white lie?"

"Yes, pretty harsh," I agreed.

"So, let me get the timeline straight, Dad." said Stephen. "You were born two years after Saskatchewan's 1941 dry spell, which sent your dad, Grandpa Ken, to Ontario for a job with Canadian General Electric as a punch press mechanic. Is that right?"

"Yes, that's correct," I replied, smiling brightly. "Most people have no idea of the minute adjustments and precise skill required to safely operate a punch press. Nor the danger. The impact of that hole punch could relieve you of an arm so fast you'd be the last to know!" I said, laying back upon my pillows for a bit. I caught my breath.

"Back in 1941 Peterborough," I said, slowing down a bit, "with Canada still flying the Union Jack, no employee safeguards, such as today's light sensor curtain, which shuts down the machine the instant your hand strays too near the steel hole punch, existed. Back in the day, if you did not move fast enough, too bad. Some of those wartime presses weighed four tons, and the mechanic had to pay full attention each fraction of every second. A single glance away or malfunction could cost you serious injury."

"Must have been terrifying," said Stephen, rubbing his arm.

"Dad was total focus, though. Mom said Dad was not only highly skilled at operating production machines, but his technical suggestions improved both operating and safety features at the factory. And guess who his most excellent teacher was? Guess who taught the twenty-five-year-old newcomer from Saskatchewan the operating functions of that scary machine? His soon-to-be-fiancé, Hazel, a capable operator whose English mother, Lillie, immigrated to Canada as a child, with her adventurous parents. In Canada, Lillie later married Edgar Woodcock, and in 1919 my mother, your grandmother, was born in Brantford, Ontario, two hundred miles northeast of Toronto. Perhaps this factory city contributed to my mother's fearlessness around complex equipment. Just as Mom elected to name her children after social engineers of authority, Tommy Douglas and Darcy McGee, so, too, had her parents given her the name Hazel, a symbol of both protection and power.

"Dad's brother, Calvin, joined him in Peterborough. He bunked with Dad and worked as a drafting apprentice. But that was not enough for him. He also studied machine design at night school and on top of all that work, he sold clothes at Grafton's Clothing on the weekend to pay for his school fees."

"Wow, do you hear the Puritan work ethic? Busy. Ambitious. Serious. Snapping up opportunity wherever found," said Stephen.

"I see it," said Amanda, an insightful woman noting a theme. "Hard-working, self-starter types. Go-getters. I see it in Colton and Camden," she said with a smile.

—

I awaken to the delicate scent of bright crimson flowers in a Mason jar at my bedside. Saskatchewan's official flower, the Western Red Lily, grows in our lightly wooded regions and is as protected as I feel right now. "Thank you," I say to my daughter-in-law. Moments later, I'm informed that I'll be transferred to another hospital room in about an hour. Wonderful news! I have been in this ICU for what feels like forever. "Lucky, lucky me." I say it again. "Lucky me!"

My skilled nurse tells me I'm absorbing liquid nutrition well and that all my monitors report steady improvement. Unable to respond as efficiently as I would have liked, unable to articulate the question about the removal of the last encumbrance from my face, I blink.

"Good," she replies. "We'll move you in an hour, Mr. Brewster."

—

"Are you back, Stephen?" I ask. "I thought you'd left."

"It's me, Doug. It's me, Dad."

Does my heart skip a beat, or does it swell?

"Dad! Good to see you!" I say. "You're here with me!" I am so happy to see Dad. It has been awhile.

"I want to say," Dad begins, "I'm so impressed that my grandson, Stephen, has grown into such a fine, thoughtful man. I am pleased that he's carrying on our Saskatchewan prairie history. So far, it spans from 1912 to 2020. One hundred and eight years of our farm's contribution to Canada's food basket."

"Thank you, Dad. Donna and I are happy for our son, too."

It seems Dad is nestling into the soft leather visitor chair. I notice the night-duty nurse, who is consistently polite, fails to see my father sitting right there in his favourite blue shirt and brown shoes.

"You know that I met your mother during that work stint in Peterborough, during that dry spell at home. What a skilled teacher. Fearless. And what an eyebrow raiser for me to learn that her name was Hazel," my dad says. "*Hazel number two*, I instantly thought when she told me. I was incredibly pleased when, one day after making an urgent machine repair, Hazel invited me for dinner with her family.

"Now, unlike us farmers in the west, I soon learned this factory family got a paycheque every two weeks. Every Sunday, the family all sat down together after church to a good meal, usually a nice roast with gravy, mashed potatoes, and all the trimmings. The remaining days until payday meant coming and going to work and sports activities—no frills. Your mom and her family, mostly working in factories for the war effort, lived cheerfully with wartime constraints. Your mom's family enjoyed daily life and couldn't have been nicer to me. They would shake hands and give good hugs; they were very physically affectionate. At first, I felt a bit stiff, but I got used to it, even though I never came close to adopting their standard of hugs and kisses. Anyway, in the summer of 1942, having told Maurice and Hazel we'd been married a few months earlier, I had the pleasure of marrying your fine mother, Hazel Margaret Woodcock, and shortly afterward, the three of us—if you get my drift—headed home to the farm."

"I get your drift, Dad," I say.

This was quite a revelation I did not anticipate. I needed to digest its meaning, but this was not the time. I'd never heard Dad dig into a conversation so enthusiastically. I decided to take full advantage, even from my horizontal position.

"You're making me think of the early days, Doug, when I first brought your mom out to the west. It was not easy for my young bride at my parent's place. Here she was, a city girl like your beautiful Donna, accustomed to running water, electricity, trams, and easy access to the hockey games she loved, only a short stroll to the town centre. She identified as a Canadian through and through.

"By stark contrast with our American heritage, at our farm it was tall rubber boots and good luck to you. It was unrelenting farm work. Remember that my dad, while half blind, had a wife, Hazel number one, who twice made up for her husband's lack of vision. I am talking about my own mother here, Doug. She began each day by issuing orders to her young adult farm workers who she employed and boarded. It would have been easy to mistake your mom for one of her helpers though, because she was the same age as them. Even with a new baby in tow, she was assigned hefty garden and kitchen chores. I always wondered how your mom contrasted her warmly welcoming Peterborough family lifestyle against this far more practical farm lifestyle. Maybe my fledgling spouse felt hurt that she was treated as a farm worker?

"I guess my parents, as was common, were all about daily life on the farm. Maybe that did not go over so well. In any case, as you know, Doug," Dad said leaning back, "you saved the day for us all."

Dad shifts the armchair even closer to me. Lightly, he touches my hand, the one clinging to the steel bar of my hospital bed. "I want to say how glad I am that your surgery went so well, Doug. Awfully glad about it."

"Thanks, Dad."

"Here you are, sitting up on the side of your bed only forty-eight hours later."

"I'm also truly amazed, Dad."

"This whole business has caused me to do a lot of reflecting, son. I have been thinking about you every day, even more so since you got affected with this heart business. I've been thinking about your life. I've also reviewed my own."

"I'd be glad to hear your thoughts, Dad," I said, relaxing deeper into the soft pillows that raise me to face my father.

"Well, Doug, you may or may not remember a summer's day way back when you were about eleven years old. Oh, I do not recall what big invoice we got, but I know it was around the time our financial situation suffered a terrible loss from the grain disease called rust. Your mother and I—well, mostly me—were worried about money. It was okay to talk about it out loud, I suppose, but then I got onto complaining about how my father, your grandfather Maurice, nearly lost the farm for us during the Great Depression of 1929, what with his wild, worldly scheme of future grain contracting. If only he would have hunkered down with what he had, been satisfied with his lot, then we'd have been so much better off. But no, years later we still owed the bank for that giant financial loss! At the time, I had to listen carefully five times to figure out what the heck 'margining' meant and how the failed finances of people I'd never met, as far off as New York, managed to swallow Grandpa and Gramma's money on their Saskatchewan farm.

"Anyhow, that evening at the supper table, your mother laid a hand on my arm and asked me to stop talking about it in front of you.

'That's Doug's grandfather you're talking about, Ken,' Hazel said. 'And he's been good to him. So, let's let sleeping dogs lie for the moment.'

"Your mother's blue eyes flashed with a tiny glint of moisture, and I knew she meant business. So, I stopped my griping about a good, well-intentioned man who had made an honest mistake. Your mother put an end to my criticism that day."

"So, no harm done," I said, from my bed.

"That is right son, no harm done, and maybe a whole lot of good. It's only now I'm linking the timeline between my anger at my father's costly bank venture and your successful ones. It's only now I'm connecting the dots from that time I complained about my dad's misfortune to you starting up your dog breeding business."

"You made the crates for me, Dad. Solid and airy. I was glad for them! You helped me ship Collie pups by train all over Saskatchewan."

"I think, son, that with my grumbling about lost money, you decided to sail us forward, way ahead of your time. I mean, only a week later, your mother told me you had an after-school job cleaning the floors at school, on top of all your chores at home.

"Now, do not get me wrong, son, I was proud of your early efforts. I was impressed, and although I had no trouble opening my mouth to complain about my father's mistake, seems it clamped shut when it came time to praise my children. Of course, I was still figuring you'd eventually work the farm with me. I had no idea your mother and I were witnessing the swift start of a genuine entrepreneurial spirit—an unstoppable one. Your mom told me you were fast skating to many new ventures. Neither of us knew, though, how fast you'd speed ahead."

"Thanks Dad," I said, taking a deep breath.

"Which leads me to some further thinking, especially with your heart issues. I wonder if maybe there was too much on your plate back then, not only the school work you enjoyed no more than I ever did, and then chores, but also the pressure from my humourless griping away about the cost of a paltry cup of gas so you could see a movie in town. I've been wondering if your overhearing me at least a half dozen times steered you to overdo it a bit."

"Well, that's kind of you, Dad, to ponder my childhood like this. But look, I never missed a meal. I had warm clothes on my back. You took your time and were a good teacher. I'm not complaining, Dad."

"I know you're not son. I know you would be very reluctant, unlike your old man, to start griping. Still, your sweet sister, Donna, your dear mother, and I discussed this the night of your surgery—that from where we are now, we agree that all of us, our whole family, maybe missed out on what we didn't know much about. After all, it's hard to miss a habit you don't practice."

"It's like this, Doug," said another, much softer voice beside Dad. I see now that the second of the three easy chairs in my hospital room is occupied.

"Donna?" I ask my sister, "is that you?" My heart leaps at the sound of my life-long supporter, encourager, and friend, a sister for whom my appreciation continues to grow each year of my life.

"Yes, it's me, Doug, and Mom will join us shortly."

A gentle golden atmosphere glows throughout the room. My gaze fastens upon my younger sister, Donna, whose brown eyes precisely focus the way Mom's did when she wanted to make a point.

"It's like this, Doug," Donna repeated. "From where we sit now, and from the view we have of the past, believe me, we're all children that contribute to our community and country and made something of ourselves. We're proud we all rolled up our sleeves and plunged into the work at hand on the farm, which there's still no shortage, as both my hard-working children Trevor and Tawnya know only too well." Donna took a breath and added, "We can both be plenty proud of our devoted farm kids, eh Doug?"

"Yes, indeed, Donna Grace. We did good!"

"But to get to the point, Doug, from our precious perch now, we are all seeing that we might have added an alternate use to those long, strong arms of ours. Might we have wrapped them around each other from time to time? Today, we all hug and sail kisses and pat each other on the back. Even Gramma Hazel is into it. She has no suggestions of how her daughter in-law-might more efficiently employ her time or improve her cooking skills. Gramma's regularly hugs Mom and tells her what a great friend and caregiver she is, and how lucky her Kenny was to marry such an adaptable city slicker who learned to live the farm life so well. Gramma Brewster's all compliments now!"

I observe my father and Donna Grace, my beloved sister—a summer baby born in August 1945. Together, we rest in the warm suffused light by my bedside. I feel their heartfelt embrace and once again enter a soft slumber.

"I'm still here, son," says Dad when I again awaken. Two nurses are leaving the room, tubes jostling noisily, a monitor beeping, and a cart of steel plates clanking down the hallway. Once again, no one said hello to Dad. I look over.

"Dad?"

"Yes, son. I am right here beside you. I've more thoughts, Doug."

"Sure, Dad."

"As you know, I now have plenty of time to chat with my parents, the hardworking Hazel who could, and did, bulldoze herself and her helpers into performing two days work in one, in addition to cooking up magnificent meals for Dad, Calvin, and I. My mom may have rolled all the unspoken affection she felt for us into flaky pastry pie crusts, large molasses cookies, and delicious milk gravy. As for my dad, he never stopped dreaming about being financially secure for his family or worrying about farming risks. He says that he is proud of what you have accomplished."

"Thank you, Dad. Thank you for saying that. I appreciate it. But what reflections do you have of your own life, Dad? I'd love to hear them," I said, nestling into my pillows to listen to a voice I'd sincerely missed.

"Well, my parents, both Americans from Montana, arrived at their Saskatchewan homestead in 1914, two years before I was born. The house is still there, five miles southwest of Earl Grey. Your grandad, who had scouted ahead of Mom to survey the property, said his first shock was to find a family already living on his property, his quarter section occupied. Still, warmly invited to supper that night, he tucked into—you guessed it—a mountain of mashed potatoes, a hill of buttered carrots, two cobs of boiled corn, and a thick chunk of cured ham. Hungry as a Montana grizzly, he still remembers that welcome meal. They sorted the land issue out over their plates, it turned out that their hosts had mistakenly settled on the wrong quarter of the four sections allotted, a common error in those days. Tracing the official map he had been given, Grandpa pointed to the four quarters, aptly named the SW, SE, NW and NE, each one comprising 160 acres for a total of 460 acres per single section. Grandpa was glad to see his house already built, and of course, he helped his new friends construct their own home.

"'It's not the Mayflower II,' your grandpa Maurice wrote to his Hazel, a reference to the ship that sailed his Brewster forbears to Plymouth, Massachusetts. Safely landed, the brave and adventurous English newcomers evenly migrated to Ada, Kansas. Sometime soon, Doug, I will ask Mary and William Brewster what it was like on that ship crossing the Atlantic in 1652.

"Now, on your gramma Hazel's side," Dad continued, "she was born on December 5, 1880, the only child of Lillie Walwrath and Willis Flint, also from Montana. Your gramma's parents married on October 6, 1886, in the appropriately named Union Springs, Minnesota. I wonder if they smiled about that. In later years, Hazel and Maurice welcomed Hazel's Mom to Canada, living in Earl Grey for a few years.

"You know, Doug, I was born thirteen years before my Grandpa Willis Flint died and before the bad financial crash on October 29, 1929. Everybody who was alive during that time remembers where they were on that desperate day. Several factors shaped my personality, Doug. I won't talk about it now, but I've changed, son."

"You have, Dad?

"Yes, Doug, gratefully so. I am not talking about my character. I did not spend a long time confessing my sins at the pearly gates. I am glad I had a good, decent character. My grain sifter taught me a lot: take the good, leave the rest alone, maybe to use as fertilizer for spring planting. Doing the right thing and helping my neighbours was a given. The Brandt's and Fiessel's helped me a great deal, and I hope I helped them too. My sins, if you would like to call them that, were not of commission. They were of omission—the things I didn't do well."

"Like what, Dad?" I asked as best I could with that last tube in my mouth.

"Well, Doug, do you recall getting any great big bear hugs from me? Do you recall me patting you on the back and telling you what an earnest effort you applied to every chore you laid hand to? Did I congratulate you on your after-school job venture or your dog-breeding business, which fetched a pretty penny at the time? Did I acknowledge that you were, as your mother said, ahead of your time? Let me answer that for you, Doug, because your mouth is full of more tubes circulating than a Ferris wheel at the Calgary Stampede. The answer is no. You did not hear warm tones or receive physical demonstrations of affection from me, Doug. I am here today to explain myself, and no, your mother and sister did not send me. I am here on my own.

"I suppose I could have learned a lot more than I did from the wonderful welcome I got to 57 Aylmer Street in Peterborough— the address your mom gave me when she invited me to meet her friendly folks and three sisters. Her two brothers were in the forces. Your mom grew up not far from Peterborough's famous hydraulic lift lock, the highest hydraulic boat lift in the world, on the Trent Canal, in a city with conveniences such as ice, milk, and electricity. She had fun with her close family, all of whom loved their community sports. And who did she get stuck with? A somber farmer with fear of poverty. The stamp of the dust bowl disaster rarely lifted, never mind the lift lock down the road."

I listened with great respect, as I do to anyone who has the courage to introspect.

"Oh, and about your earlier question, son, the reason we never separated the cream from the milk and sold the cream was that is was easier not to. Tremendous time and expensive equipment are required to separate the cream. The pressure to get our fresh milk to the Gibbs train on time, not soured, was enough. Delay meant loss. Now, goodnight son," said Dad. I felt his hand on my shoulder as I fell back asleep.

—

"We're checking you out of this luxury hotel, Doug. Enough time off, wouldn't you say?" said Dr. Dory with a laugh. "On Monday, you're on your way home."

"Excellent, Doctor. I don't know how to thank you."

"Well, follow our recovery program, and you'll soon be a productive citizen again."

That evening, Donna, my wife and best friend, had a gift for me: my collection of memoir notes, along with my favourite pen. "Thought you might like to review the notes you have, Doug, and then when we get home, maybe add more."

"Can you hand me my shirt, please?" I asked Donna, eager to remove my blue hospital gown.

"On Monday, Doug. We've still three days to gather our strength," said my wonderful wife. With Donna, everything is "we." "How about, for starters, we practice getting you in and out of bed?"

It was enough for a Monday morning, and plenty of effort for that afternoon, too. That evening, my hospital bed was raised to a fully seated position, a real graduation. One machine, the tick-tock thrummer, was detached and rolled out of my room. My children's interest in our family history, and their appreciation of how good Canada turned out to be for all of us, gets me to reflect even more.

—

As Dad recounts it, my grandpa Maurice Brewster arrived in Gibbs to claim his homestead in 1913. He had travelled direct on the CPR line from Montana to Saskatchewan, then he unloaded and transported his worldly belongings some eight miles to his Earl Grey property. Thinking of his belongings packed into a grain car reminds me of the sled box that carried me to my first home. I wonder if this mobile theme, be it on sleigh runners, train tracks, wheels, or floating on water perhaps played into my later establishment of Brewster Trailer Manufacturing, manufacturing mobile homes? It launches my successful business career, which I began back in the 1960's when I was in my twenties.

—

"I recently heard tell of one of the first adventures of our forebears, Doug," said Dad. As Hazel played and sang her favourite tunes last evening, our forebears, William and Mary, listed for us the provisions they brought from England to Holland, and from there to Plymouth Rock on the Mayflower in 1612."

"Holland?" I ask. I glance over to see Grandpa Maurice comfortably seated in the high-backed visitor chair, one eye beaming with pleasure as he recites the list of goods carried onto the Mayflower by his fore parents, William and Mary Brewster. The couple, threatened in England by their choice of faith, escaped by travelling to Holland, another port of departure for the new world.

"And I recall," said Grandma Hazel, "their deep dedication about which you so inspiringly shared with me in your wonderful proposal letter, Maurice. Your family's history of commitment was one of the reasons I accepted your invitation to a life together."

"You know, Hazel, in 1921, after we were married a few years and our Kenneth was still young, I learned that the Reverend William Brewster designed and wrote the Puritan Compact while on board the Mayflower. On November 21, 1692, on the deck of that vessel, he and ten other men signed his composition.

"I'm so glad you said yes, Hazel." Maurice turned to the woman who had seen him through forty years of tough-as-nails farm work.

"Despite the 350 years since John Calvin's religious philosophy took hold," said Hazel, "it still makes sense to me. I was proud to name one son Calvin, after the inspiring Puritan, and if I had had a girl, I'd have called her Mary. That couple meant business, and so did we. Would you like to hear the tale as told to me by Mary Brewster?" asked Hazel, happy as a lark. Then she began.

"Her story of the Mayflower and Plymouth Rock tells of the formation of a little congregational church at Scrooby, North

Nottinghamshire England, a separatist church cradled in the Bishop of York Brewster Manor, home of Mary and William Brewster. Their heresy cost them a hefty fine and a cold, damp, winter imprisonment for William. Still, risking his life for this Christian church, William secreted his little flock to Holland, where his duty as elder was to build up and administer their new church, preserving its doctrine.

"William paid every expense of every member of his church, especially those who'd left their sources of income to flee to Holland. That William composed and drafted the Puritan Compact of November 21, 1620 in the cabin of the Mayflower seems certain. He was the moral, religious, and spiritual leader of the Colony during its first years of peril, its Chief Civil Advisor and trusted guide until the time of his death. If not for the Compact prohibiting a minister holding a political position, he'd have governed the colony. The signed document of the Compact testified to the travellers' promise to be fine examples of Christianity in the new world, to honour the country in all their affairs, and to be of service to and Christianize the Indian tribes.

"Mary Brewster can still recite their provisions list," said Hazel. "Canvas sheets bolstered with dry straw, wool rugs, and blankets—because Mary believes that good sleep prevents illness and preserves health. Enough food and water for the daring journey across the Atlantic," Hazel continued, "with more to spare to give a fellow traveller a little extra support. Crates of biscuit, oats, peas, wheat, butter, codfish, vinegar, and rice loaded on board. And, oh, yes, a great variety of seeds. Lots and lots of seeds!

"Of course," nodded Maurice.

"Mary says that while William packed arms, priority was given to utensils. She showed me her beautifully handwritten page listing, in order of importance, iron pots, kettles, frying pans, wooden spoons, a mortar and pestle, soap, farming tools—such as hoes, axes, handsaws, hammers, shovels, spades, augers, chisels, gimlets, hatchets, grinding stone, and nails—and finally, locks for the doors. She said they brought a bare minimum of clothing, letting the seasons be their guide, not fashion or flare."

—

I ask, "Why did you name Dad Kenneth Willis, Grandpa?"

"We honoured Hazel's father, traditionally, by giving our first-born his name, Willis. As for Kenneth, it's a widely known, practical name, suitable for a boy who would be meeting international neighbours in the west. Kenneth is a name claimed as their own by the Scots, the Irish, and the

English. It is easy to pronounce and means handsome and born of fire," answered my grandpa Maurice.

"And a second possible answer," added Ken's mother, Hazel, "is that one famous bearer of the name was the British novelist Kenneth Grahame, who wrote *The Wind in the Willows*. It is a classic children's story, filled with adventure, morality, and wonderful tales of friendship between three animal friends. One day, after trying on a beautiful flowery hat I made for her in Berkley, California, one of my satisfied customers gave the book to me as a thank-you gift. She loved my stiff paneled hat, a design embedded with tiny dried flowers and finely cut glass, and I so enjoyed the storybook!

"One evening, five years later, there I was, married to Maurice, living on a homestead in Saskatchewan, and the mother of two sons, Ken and Calvin. As I read to the boys one winter afternoon, I shared that this author, also named Ken, wrote about forever friends who worked hard to solve their problems and remain friends."

"I'm happy, Doug," said Grandma Hazel, again visiting me. "So glad to have the time, at long last free of chores, to while away the hours and share our family history with you."

"Gramma, I'll never meet these good people. When I hear about our forebears who I never knew, I feel I've missed out and now I'm playing catch up."

"But, Doug," said Grandma Hazel, "you didn't miss out on a thing! It was not your turn to be alive, dear. We each live in our own time and place. It does not matter that you did not shake hands with those who lived before you. That is not the point. The point is that each of them did their part, willingly and devotedly. And their blood runs in your veins, minute by minute, as you listen to their stories from which you can, if you wish, compare their values to yours. There are thousands of critical moments in your history when you choose to honour your life. That is when you discover if your values match or differ from those who preceded you. Such choice moments, Doug, are when you advance your family history down the same road or onto a new path." Gramma Hazel looked me straight in the eyes and said, "I'm glad I designed my hats. And as for never getting a chance to meet our forebears, well, you and I are visiting together right now. What does that tell you?"

"I'm happy to listen, Gramma," I said, settling in to hear her eager recitation. I paid rapt attention, not as a kindly favour to a senior relative in my hospital room, but as a grateful human being, refreshed to be part of such a purposeful, energetic family. "Please tell me about your parents, Gramma."

"Well, Lillie, my mother, was born December 5, 1860, in Carson, Minnesota," said Gramma, smoothing her silk gown.

"She was laid to rest in Mile City, Montana on March 22, 1955, next to Willis, her husband—my dad—who had passed away in the early 1930s. Mom was in an excellent retirement home for many years in Miles City, where we spent winters and warmer weather visiting her. As for my dad, I have a poem he wrote before he passed away, which reflects his early Montana life. I would like to read it to you:

GOOD BYE TO MONTANA
by Willis C. Flint around 1930

I have ridden your hills and valleys,
I have camped on many a creek,
I've tried my best in every way
To turn the cowboy trick.

But time is passing swiftly
And I am growing old,
And my thoughts go back to my boyhood
And my dear old Mohawk home.

Yes, I've rode your hills and valleys,
I've rode your glorious plains,
I've put in many a pleasant hour,

My hair is turning silver grey,
my eyes are growing dim,
but when I think of the dear old home,
I feel like a boy again.

Then good-bye to the hills and valleys,
Good-bye to the Golden West,
I am going back to the Mohawk,
To take a little rest.

To ride with a bucking bronco,
or to rope a vicious steer,
May be a trifle for a young man

With my fleet foot Gypsy Maid.

But my thoughts go back to the Mohawk,
And the pleasant books and hills,
And I long to feast my eyes once more,
On the dear old Glenville Hills

So, good-bye to your hills and mountains,
Good-bye to your bucking broncs
Good-bye to the boys with the spurs and chaps,
and the girls with the curls and pumps.

I am going back to the Mohawk,
Never more to roam,
And when I cross the great divide
It shall be from the dear old home. I But not for one of sixty years.

So good-bye to the land of the prairie dog,
The coyote and moonshine booze,
To the land where you wait for your next years crop,
To buy your baby shoes.

I am going back to the Mohawk,
To the land of the apple and the rose,
To the land of the sugar maple
To the land where the chestnut grows.

And if per chance you ride that way,
Just give the country yell
You will always find a welcome
If you call on Uncle Bill.

"While jolted by the Great Depression, my mother enjoyed another thirty-five years with us and her precious sewing circle family of friends. I loved visiting those women to remind them how much they mean to me," Hazel finished up.

"It's my unsolicited opinion," said Grandma Hazel, "that your Grandpa was ahead of his time in looking forward to the future. Margining his crop to earn more money was a risk he took. I'm proud of him for it. Perhaps Maurice's brother, your great uncle Callie from Britt, Iowa, influenced his decision. I'm not sure. Callie, a family doctor, was quite involved in stocks and perhaps played the margins, too. He kindly helped your grandpa with a $100.00 loan to purchase our original Saskatchewan lands. It all worked fine until the Great Depression of 1929 hit, along with the deadly 1930s drought and the Second World War in the 1940s. Those were grim times, Doug."

The laughter in Gramma Hazel's voice faded. "The burning desiccation of every living plant on the farm, Doug, the stark dehydration of life that cracked and shriveled the ground beneath our feet, was a shock. I'd never seen the likes of it in all my lifetime. Here we were, with two young boys, farming in a dustbowl of greedy grasshoppers. A world war so close on the worn-down heels of the first, which shaped your Dad's youth."

"I'd like to hear more about it, Gramma."

"Well, in the late 1930s, when you dad was in his early twenties, he left the farm during winter months, to work as a carpenter in Britt, Iowa with his uncle Callie, who was building a new medical clinic. Self-taught, our Ken was a better carpenter than most, I would say. To me, the humble ones who doubt their abilities often work far harder than those overtly proud of their paper attestations. Your dad perfected constructing farm buildings and repairing his equipment, just as you have done, Doug. I wonder if you were inspired by your father's longing for proper equipment. Was this your motivation for establishing an Earl Grey company to provide easy rental access to essential farm equipment for farmers in your community?

"Honestly, back in that time of poverty in the early 1940s,

I was glad he had three squares a day in Iowa, thankful he was able to support himself. Your dad was not the self-congratulatory type, but he grew to allow himself some measure of pride in his achievements. I was glad that Ken's flat feet exempted him from the armed forces during the Second World War, because my oh my, believe me, Doug, we desperately needed him on the farm during the summer growing seasons. For sure and certain, like his father before him, your father worked like a maniac. Relentless in doing his part, he was persistent in keeping the farm equipment in good repair, so they would operate better in our field.

GIBBS FARM SALE AND ONTARIO MOVE

I LAY QUIET, FLOODED MEMORIES OF MY farm life. I did not overhear my parents' forceful conversations in the kitchen long after I fell asleep in my farm bedroom. I learned only later of the intense talks that resulted in the sudden auction of all our goods and farmland before our 1948 cross-country move. I wasn't privy to the depth of emotion contained (or not) in a decision that soon had me pressing my nose against the window of an extensive mobile conveyance, with box cars riding along rails much larger than the horse-drawn ones sliding along snowy fields.

I treasure the misspelled "Moday Farm Sale and Auction" poster, the one listing the land, house, farm, all of Dad's slowly acquired machinery and tools, along with every bed, table and chair, farm attire, and every sack of carrots and potatoes in the storage cellar. The flyer for the Monday auction, posted at Heckman's general store in town, at the post office, and the local barber shop, welcomed all buyers to the usual delicious free lunch—a United Church fundraiser that doubled as a farewell for our family.

AUCTION SALE

— OF —

Dairy Cattle, Power Machinery, Furniture

Sec. 15-22-21, ½ mile East & 2¼ miles South of Gibbs; 6 miles West and 5½ miles South of Earl Grey; 2¼ miles North and 1 mile East of Wilde's Filling Station

— ON —

MODAY NOV. 8

Dairy Cows

Red Cow, 5 yrs old, milking, to freshen in Jan
Holstein Cow, 7 yrs, milking, to freshen March
Roan Cow, 8 yrs, milking, to freshen in April
Holstein Cow, 8 yrs, milking, freshen in Feb
Black Cow, 8 yrs, freshened on Oct. 26
Red Cow, 9 yrs old, milking
Black Cow, 8 yrs, milking, to freshen in March
Red Cow, 8 yrs old, milking
Black Cow, 7 yrs, milking, to freshen in March
Roan Cow, 8 yrs old, to freshen in Nov.
Red Cow, 7 yrs old, to freshen in Nov.
Heifer, 3 yrs old, to freshen in Jan.
Heifer, 3 yrs old   Heifer, 2 yrs   Steer, 2 yrs
Heifer, 1 yr   2 Pigs   50 odd Hens

BUILDINGS
Smoke House 6 x 8   Granary 12 x 14
Brooder house 8 x 10

Farm Machinery

I.H.C. Farm All M Tractor, good shape, new rubbers
3 furrow John Deere Tractor Plow   Hay Rack
10 sec. Diamond Harrows with rolling drawbar, new
21 ft I.H.C. Disc, new last spring   Hay Rake
5 sec Spring Tooth Harrow, new last spring
24 run s.d. John Deere Drill on rubber, new last spring
12 6 I.H.C. Stiff Tooth Cultivator   Manure Boat
8 ft I.H.C. Binder   Steel Truck Wagon
Wood Wagon Gear   Rubber tired 4 wheel Trailer
2 h.p. Fairbanks-Morse gas Engine, new last spring
Challenger Pump Jack, oil bath   Wagon Box
Steel Stock Tank   Tack Heater   Paige Wire
320 rods Barb Wire   Steel Caboose Trucks
16 loads Sawed Wood   125 Treated Fence Posts
2 sets Sleighs   Steel Circular Saw Frame, 2 Blades
Jamesway Oil Brooder and hover, nearly new
Set Work harness and collars   3 Water Barrels

Furniture

Viking Power Washer, Briggs-Stratton motor
Oak Dining Suite; Buffet, Table and 6 Chairs
Davenport   4½ ft Bed, complete
Cot   2 Cribs   2 Ranges   Kitchen Table
2 Kitchen chairs   Commode   Dresser
Wicker Rocking chair   Dinner Set for 6
6 tube Philco cabinet Radio   Copper boiler
Coleman Gas Lamp and Iron   Canned Fruit
Sealers, Pots, Pans, Etc
Kiddies' Set, 2 chairs, 2 Rockers and Table

Miscellaneous

32 BAGS OF POTATOES   100 LBS CARROTS
STACK OF HAY   2 STACKS OF OAT SHEAVES
PRESSURE TIRE PUMP   18 MILK CANS, 8 GAL SIZE
DAIRY PAILS & STRAINERS   HORSESHOE PULL JAW
DEHORNERS   NEW DODGE HALF WASH TUB
GRINDER STONE   WIRE STRETCHERS   SLOW SAWS
SOFT MAPLE, ASH, CHARMS, FORKS, SHOVELS, FORKS
AND OTHER ARTICLES

Sale Starts at 10 a.m. Sharp   Lunch Served

TERMS CASH

Ken Brewster,
Owner.

C. F. Bauer,
Clerk.

Bob Bauer, Auctioneer,
License No. 5181

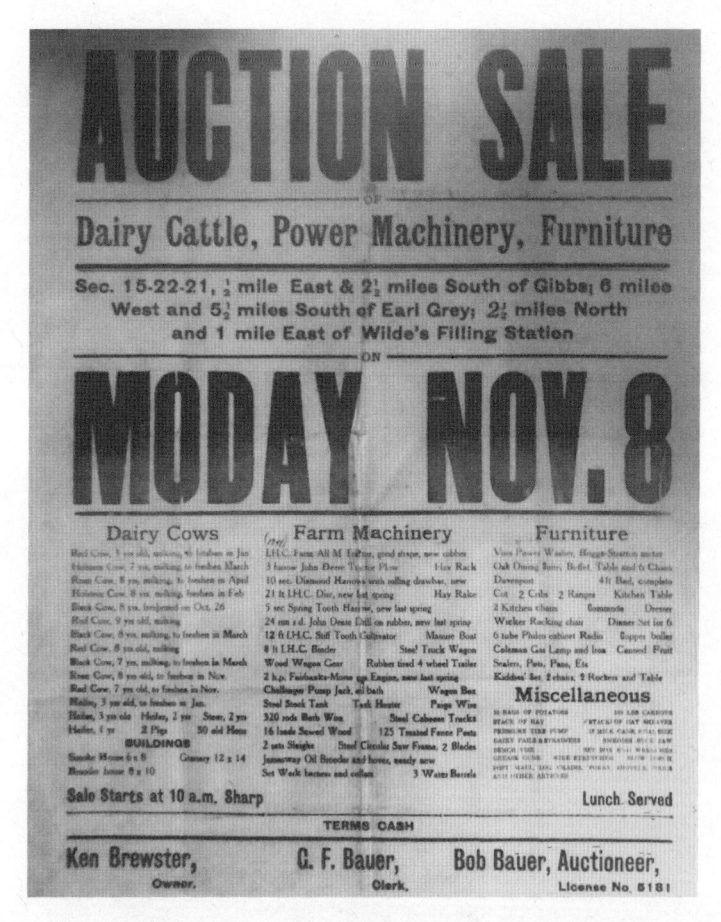

I counted sixteen cords of Dad's sawed wood, his wire stretcher, and lengths of barb wire, as well as one hundred pounds of potatoes and carrots from the basement, all snapped up by a farmer with a slight sign of his index finger. I saw townsfolk in small groups watching each other like hawks before a single nod, an almost imperceptible gesture not lost on the auctioneer, signaled a sale. Mom, her face beaming with happiness, handed me a thick egg salad sandwich for one hand and an oatmeal raisin cookie for the other. The local United Church cooks, experts at fund-raising, served up delicious, steaming-hot home cooking and baking.

"We're off, Doug! We're off on an exciting adventure!" said Mom as I knelt at the window of a real train. Handing me another oatmeal cookie studded with raisins, Mom beamed her appreciation at Dad, whose gaze fixed toward the same window as mine. Silent, with a solemn expression, he turned to face Mom. With our packed steamer trunk pre-sent to Peterborough and our luggage pre-checked, we chugged out of Gibbs. Grandpa Maurice dabbed one watery eye with a handkerchief the instant the train whistle blasted our departure, and Grandma Hazel hugged us all before handing Mom a generous bag of her oatmeal raisin cookies and thick baked ham sandwiches.

"Slathered with yellow mustard, the kind I ate as a child in Montana, Hazel." Grandma went on and on, not knowing how to hold the grief she felt so suddenly. "You know, Hazel, the states of California and Montana once monopolized mustard production, yet Canadian acreages increased it because of the fertile soil. No sandwich escapes my spatula's smack of the best hot spice in the west! Goodbye, Hazel. Goodbye, Ken. Goodbye my darling grandchildren."

"We're done," sighed Mom, offering a cookie to Donna. "We're done with inhuman claims on my good husband's every single moment, outrageous demands made by animals and buildings, and crops that compete for attention 24/7. Goodbye to the gobbling up of a young farmer's life, a parent with two kids, a father who's every moment of his every day is consumed with the feeding, stoking, cleaning, driving, seeding, and harvesting, never mind a moment or two with his family. Goodbye to our dependence upon elements that may, at their whim, destroy us with unforeseen drought or torrential rain or brutal insect infestation. Goodbye to back-breaking insurances, seed, tax, feed, heat, and machinery payment bills that drain us so thoroughly we need to calculate the cost of gas for every trip to town!

"Instead, Kenny," sang Mom, her voice passionate as she leaned over and placed a warm hand on Dad's knee, "its 'Hello!' Hello to steady, guaranteed paycheques every two weeks, regardless of the weather's whims. Hello to two weeks of paid vacation every year. Hello to regular evening supper with your family, getting to know your children before they've grown and gone instead of repairing machinery all night in the barn, hello to having a little fun every week, and hello to spending more time with the woman you married not so long ago." Mom smiled beautifully, her face aglow.

"Hello, hello!" Little Donna mimicked her mother, causing all four of us to smile.

"We're on our way, Hazel," said Dad, serious and somber, perhaps a tad terrified. "Glad we cut even, all bills paid up and enough to start up anew."

That fall of 1948, Dad had turned around and sold our farm. What happened? How come?

"Mom, I want to see Grandpa Maurice. Where's Gramma Hazel?" I'd ask.

"We'll visit, Doug." I would be told. "We'll see them soon.

"So long western Canada! Goodbye dawn-to-dusk exhaustion, hard work, low pay, no electricity, and never enough," said Mom, Donna cuddled in her arms as the train chugged its deliberate way out of the Regina elegant train station, with its very high ceilings and long wood passenger benches.

"Bye," said Donna, curling up for a rare nap while I watched Dad tally a column of numbers on his notepad.

The warm welcome at the Peterborough train station was a thrill. Hugs, red and white balloons, yelps of laughter, kisses, and smiles galore filled the station. It was quite a welcome for our young family from rural Saskatchewan, though, eating freshly baked cookies while waiting for our baggage to arrive. A wrapped gift from my uncle Bill was slid into my little hand. Wow, a toy tractor! When Dad saw it, a shadow grazed his face, as though he had glimpsed his precious machine, freshly sold off.

—

"Doug, Doug, its Christmas! Gifts are waiting for you! Donna's downstairs sipping hot chocolate topped with marshmallows." The Woodcock family clan lives happily under their one roof of 57 Aylmer Street. My sister and I are their most welcome newcomers. Mom's family of nine get along so astonishingly well; they are always chatting and joking and coming and going out together.
I like it here, with all the city convinces, such as horse-drawn milk and ice carts with home delivery, bathtubs with running water, and electrical lights. I thoroughly enjoyed lying in bed, at night, listening to the street traffic noise outside my window. Really!

My second walk with Dad to the red brick building, King George School, was just a few blocks away and no less exciting than the first practice visit the night before, when he showed me the public school where I'd learn to read. "It's always a good idea to verify an address ahead of time, Doug. That way you will not get lost at the last minute," said Dad. My formal education began!

I am sure I was quite uncomfortable starting in the middle of the school year and not knowing anyone.

My parents were settling into their surroundings, Mom ecstatic, I am sure. One morning, as Dad lifted a full coffee cup to his mouth, he quietly announced his plans to buy a grocery store, Brewster Food Market, but only after he has spent a year or so learning the business in a grocery store operation. "I'll apprentice in a store called Loblaws—train on the job," he said.

"Perfect, Ken," Mom said, raising her cup in a toast.

Mom announced she would take Donna and me to a hockey game the next night, rosy-excited with plans. On Monday during an afternoon tea with Mom's old school friends and my sister, I was at school meeting my new classmates. With no cows to milk at 5:00 a.m., no horses to tie to hay wagons, no sloppy cement floors to hose down even in winter, no machines to repair, Dad attended his first and only job interview. He was hired on the spot.

"Enjoy yourself today, Hazel," Dad said, heading out.

Mom enjoyed the ease of her friendships, the city, and fun sports outings with her brothers and sisters. Donna and I were fussed over by our animated Woodcock aunts and uncles.

—

Why, then, did I see my freckled six-year-old face mirrored in yet another moving vehicle, this time a car, in the early summer of 1949? Was I headed back to farming in Saskatchewan? The story goes that Dad had not had the amount of money needed to buy a grocery store. Perhaps over the few months in the big city, he came to the awareness that the prairies and farming were his best option. I expect Mom was not overly excited, but she accepted the reality of the situation. I heard tell that Dad felt lost in a sprawling anonymous workforce; however, for years afterward, Dad's practiced the art of properly arranging cardboard boxes inside each other to save space—a skill he learned from his experience working at the grocery store.

And Mom, did she notice the difference in her brothers? They were army vets who were formerly so excited about life but now affected by the trauma they had suffered overseas. The Woodcock family had increased in number, as well as new personalities. Things had changed during the period 1943-1948. Did some family members resent that Dad had not endured the same ordeal as her brothers? Perhaps those who had no idea of the rigorous demands and dangers of farm life felt this way. Perhaps after noting a goodly number of changes herself, Mom reconsidered the advantages of living in the west?

The Woodcock family, truly generous in embracing our family, left a permanent impression on me. Donna and I knew we belonged and would be loved by them forever. We would continue to visit and be visited by them over the years, growing up with many good memories of our eastern grandparents, aunts, uncles, and cousins.

"We're heading back!" chanted Mom from the front seat.

"Goodbye, Ontario," I said, kneeling on the back seat to face the window. Slowly, Peterborough disappeared. I sat down and faced forward.

For the trip home to the prairies, Dad purchased a beautiful black 1949 Plymouth Sedan and a small, single-room silver travel trailer. He also promised Mom attractive new furniture for the house we would soon have. I did not finish my year at King George School, because we hurried back to Saskatchewan for the spring seeding. I thought about the dear kindergarten friends I'd always remember. "Bye, Timmy, Neil, and all my cousins. I will miss you," I whispered.

The small silver trailer, a ten-foot long unit, smelled of the stove's propane gas. It carried us first to the Great Lakes, where we ferried from Port Arthur to Thunder Bay—Dad's favourite geography of his travels anywhere.

GROWING UP IN EARL GREY, SASKATCHEWAN

AT THE TIME OF MY BIRTH, WE were three people—Dad, Mom, and newborn me nestled in with my grandparents. Now we numbered four and, upon arrival, set up our little trailer just south of the family farmhouse. Uncle Calvin and Aunt Jean (Laurie) were just married and living there with Grandpa and Grandma, too, but they were looking to move to Earl Grey. Uncle Calvin decided to start a business clearing bush land and purchased a new International TD-9 Crawler tractor that summer, and I got my first caterpillar ride with him when he drove it home from Gillespie's Dealership in Earl Grey.

By today's standards, it was small and slow, with no hydraulics for lifting attachments but lots of traction at slow speeds. It had a newly designed diesel engine that could start with a hand crank, when the electric starter failed, and initially and started on gasoline. I was impressed and have continued to enjoy the rugged track machines for clearing land and bush. Uncle Calvin had a blade type tree cutter and a floating blade to pile the brush installed on the front of the cat and did a lot bush and field work for the first couple of years, with many hours spent clearing roads blocked by snow drifts for the local area.

There was a fenced area for fox raising on Grandpa's farmland. Uncle Calvin had developed it for selling fox pelts. It had a tall perimeter and solid fence with many wire pens and enclosures for each fox family. At feeding time, free meat was placed in each pen containing animals. They were quite wild and would scurry around the pen and into their shelter, but they were beautiful animals.

One hot summer day, I was helping Grandpa gather up poplar wood for cooking meals on the kitchen stove, and he told me that when my dad was a little older than I was at the time, he had ridden his horse to Semans and back, a community some thirty miles to the north. An owner of land next to Grandpa's, now living in Semans, wanted to sell it, so Dad was sent to see him with an offer. After sleeping the night in a hayloft, he secured the east half section of 19-22-20-W2, just south of their farm, for renting and later buying, before returning home.

The summer passed with Dad looking for a farm to buy. As luck would have it, the Macgregor's, Mac and Florence's farm, had come available. It was once a fine farm but now desperately needed Dad's carpentry skills to restore it to its former glory. The farm we had vacated the year before was five miles away from Grandpa and Grandma. Wonderfully, our new home was only two miles southwest of their place. It was a complete farmyard, with three-quarters of good land. We were home again. Sometimes you must step back to step forward. Now, I was six.

—

"I'll never forget my first auction," I said, my pain meds kicking in. "It took place on a sunny day in September 1949. There I was, a skinny six-year-old boy, walking the three-quarters of a mile home from my first day at Fosterdale School. Aunt Jean had taken me that morning. I will always remember that instant of joy and excitement. The auctioneer talked so fast I thought he was a magician.

"Whadilyagimme for this clean, well-maintained tractor?" he would say. The McGregor's were selling all their farm equipment, along with household items. The auction sale was winding down as I approached the yard that has remained in our family to this day. Soon, there was a great deal of commotion as the new buyers and bidders loaded up their new purchases and headed home. It soon was incredibly quiet, a vacant farmyard waiting for its new owners, the Brewster family.

—

My country school comprised of a single classroom with a grey stuccoed exterior and lots of windows to the east, for extra natural light. It did not have electrical power but was located three quarters of a mile north of our new farmyard on the corner of our land. The two acres for the school site was purchased for $40.00 per acre from Mrs. Ellen Best and her family in 1928, arranged by Joe Zimmerman who was renting the land at the time after Mr. Joe (Joseph) Pinkney's sudden death in 1916. Joseph was Ellen's first husband. The construction was completed by a local contractor and cost under $5,000.00. The required financing was obtained with a government debenture paid back by a school tax levy on the farmland and improvements in the district. I expect this was the beginning of the municipal education tax on our annual property notices.

Naming the school Fosterdale, I believe, was a significant issue. Although being unable to verify the facts, I believe "Foster" comes from the last name of a local family in the school district from Southampton, England. Frank and Georgina, with sons, Harold and Guy, and daughter, Una Gundred, had both sons join World War I. Harold was killed at Passchendaele ("dale"), France, and Guy lost his sight at Vimy Ridge. Guy returned to England, learning to read braille and educate himself. He then returned to operate the family farm. With some outside help running the equipment, Guy managed the mechanical and carpentry work himself for many years.

Fosterdale Our Families school

My dad and his younger brother, Calvin, were among the first to attend the new school in 1930, after starting their schooling in Earl Grey. The students were required to help with certain duties around the school, such as sweeping the floor, cleaning black boards, and keeping the furnace going when needed. One story goes that Uncle Calvin, during his shift, went early to light the school furnace, well before teacher Mary McDougall arrived on horseback. Once he had put her horse in the barn, he completed his other tasks before the school bell rang. When at the end of the school semester of 110 days service, little Calvin requested his pay, he was informed there was none. Apparently, he looked quite sad at this news; however, a school board member, Mr. Brantford, found a willing source and secured the $3.00 for him. Later that month, the teacher sustained a salary drop from $900 per year to free room and board. In the early 1930s, farmers could not pay their taxes. Despite this blow, being a dedicated teacher, she carried on.

The Union Jack flag flew out front every school day until it was replaced in 1965 with Canada's own flag. Just as the picture of King George VI hung on the back wall before the coronation of Queen Elizabeth in 1952. In Grade 1, the day began with the national anthem and ended with the Lord's Prayer.

The strap was the main disciplinary tool when I was in school. It was used often, although I never received it— perhaps because I was generally good and had a little luck. However, my sister, along with many others got it once or twice. A frustrated teacher extracted the strap from a desk drawer and demanded the student hold out their hand for stinging whacks. Pulling one's hand away made things worse. A couple of times, for more serious rule-breaking, the local board arrived to administer the punishment in the basement.

The school's large teaching room comprised of five rows of desks and the teacher's desk up front, with the hand bell sitting on one corner and that threatening strap in a side drawer below pencils, chalk, and paper. Grades 1-8 were taught with most subjects being combined for two grades, allowing the teacher to spend the necessary time per student with only four main groups: Grades 1–2,3–4,5–6,7–8, Grades 9 and 10 could be taken by correspondence, with assistance from the teacher and the student receiving subject material and completed assignments by postal mail.

Although common for rural schools to have a separate living space (a single room) for the teacher during the school season, Fosterdale had no such room to house the teacher. The girls' washroom was behind the desks on the left of the teacher's desk, and on the right was the school entrance with a cloakroom for coats and boots. The boys' washroom, furnace room, and coal-wood bin were downstairs alongside a large playroom.

In the winter months we played fox and geese in the snow or hopscotch in the school basement when the weather was too cold and rainy. We played anti-I-over over a little four stalled red barn roof, which housed the students' and teacher's horses.

Christmas time was extra special, with our small stage being set up and decks moved as we practised for our concert.

Soon Christmas holidays were upon us, and the community concert night arrived. The snacks and lamps to light the school were brought by the community, everyone sang songs, and students performed productions on our little stage. Soon, Santa came with gifts for all the children.

Generally, two teams of horses were tied in the four-stall red barn for the children living too far to walk. The rest of the yard was used for sports, games, and skating. One cold winter, I helped flood our rink with my favourite black pony pulling the stone boat, with a forty-five-gallon barrel of water for flooding the home-made rink, back and forth from Harry White's yard dugout. Pieces of sawed poplar wood floated in the barrel to help secure the water from sloshing out and freezing on contact with me and my stone boat. I'd hand-dipped the water out of the two-inch ice hole and into the barrel, then I'd tip the barrel of water onto the proposed skating rink. My hands were freezing in my wet heavy leather farm mittens. Like my dad and Uncle Calvin before me, I did the school chores for a few years, cleaning the chalkboard, sweeping the floor, and dusting the chalk brushes after school, earning $0.10 a day.

We usually walked to school. A ride from Dad was the exception. On very cold days, below forty degrees Fahrenheit and blustery, Dad took us on the stone boat, after first attaching the harness and horse to the stone-boat. Dad would hold the steering reins, and we would stand or sit on a bale of straw. A couple of times in Grade 1, Dad sat me behind him on our old tan workhorse named June and galloped through a blizzard in negative forty-degree Fahrenheit, dropping me off at school and riding home. Neighbours sometimes gave us a ride, too. One sympathetic father, Henry, had a daughter named Jean who was also in my class of four. He often drove us home from school, especially if it was raining or freezing cold.

Two of my Fosterdale teachers permanently resided in the community. They married farmers. Mrs. Delores Pinkney taught me in grade school. She was a great teacher and neighbour, and many years later, I had the honour of speaking at her retirement party. Aunt Jean taught for a couple of terms, along with substituting from time to time. She was a great teacher and aunt. In high school, I babysat both the Pinkney girls and my Brewster cousins Vern and Spencer.

The school closed in 1959 with Donna finishing Grade 8. She continued her education in Earl Grey. In 1963, the school and its two acres of land were sold by public tender, and the schoolhouse was converted to a six thousand-bushel grain bin and moved to our yard by Dad, returning the land to productive agriculture land once again. The windows and doors were filled in and re-stuccoed, with rods crossing every three feet to brace and hold the heavy weight on the walls when full of grain. Dad's carpentry expertise is still on display with the fine workmanship in its new role, extending its life for years to come. The little red barn got sold and moved by Marvin Smyth and converted into a car garage at, would you believe, the same farm at Gibbs where I lived for the first four years of my life.

—

We were farming again. The three quarters of a section was now Ken Brewster and family's farm. A large truckload of new furniture was purchased immediately, and we settled into our new surroundings. I do not remember how Dad found and purchased most of his farm equipment. The first pieces of equipment were always well-maintained but older, except for Dad's new Massey-Harris 44 tractor, a red forty horsepower (HP) with no cab or power steering. A new one-ton International truck was also purchased for farm and personal use. Initially it was without a steel box, due to the after effects of shortages caused by World War II.

44 MASSEY HARRIS Tractor

The day in 1950 when Dad took possession of our new tractor from Renwick's Massey-Harris dealership in Bulyea, I rode home standing between the seat and the fender. As we drove along the #20 Highway in the ditch, we collected tossed pop and beer bottles to cash in for extra spending money for Donna and I. Being my father's son, I was always on high alert. My Boy Scouts unit, a few years later, reinforced this quality in me with the motto "Be Prepared."

I was now old enough to help Mom with settling into our new home. Keeping our food from spoiling was a challenge without any cooling systems run by electricity. The city residents, like our relatives in Peterborough, had ice boxes that got ice supplied by a horse drawn delivery cart going door to door on each street. There were ice huts in rural areas, dug into the side of hills, covered with straw and sawdust, and supplied with ice in the winter. These worked for the cooling and keeping of food the better part of the summer months; however, they were few and far between. And Dad had no interest in taking the time needed to build and maintain one. Many areas had a group of local farmers that took turns supplying beef and pork to each other every month. The fresh meat could be kept in ice huts or hanging down the water well, as we did for the thirty-day period, unless a farmer milked cows and had chickens, powdered milk, eggs, and oil-based margin had to be obtained. A cool, dry, dirt basement, as our house had, filled with Mom's canned preservatives and potato pit helped greatly in our survival. Mom made bread and buns every couple of days in our kitchen's woodstove oven, the right amount of heat being regulated only the amount of wood supplied. In the summer months, the margarine was runny, and it was frozen solid during the winter months.

Dad continued to increase the farm size and income by adding land and livestock, working hard with good management. The rural road allowances were sixty inches wide, which provided excellent feed for our cattle and helped with the task of herding, preventing them from straying or delaying passing traffic. This task was our, the kids', responsibility. Donna and I would play and sometimes forget about our responsibilities, and then we would be getting in trouble for letting the cows wonder into a neighbour's field or yard.

Dad also saw the benefit of clearing brush and grass lands for added production of grain, which was usually a family job. We would pick up sticks and stones after the trees that were being cleared by hand chopping, we would pull out using the tractor and chain, and we'd clear the land plowed by Uncle Calvin's TD6 caterpillar and brush cutter. The combination of good farming and new land production was the secret to Dad's success. There were never enough grain bins at harvest with grain in piles on the ground or other temporary facilities. In 1953, our yard next to our house had a round enclosure, about sixty inches in diameter, made with posts pounded in the ground and wire attached. The inside was lined with oat sheaves to hold the fifteen hundred bushels of wheat. Dad understood the importance of fertile land–uncultivated land versus sixty years of cultivation. It is too bad Dad did not live to see the future agronomy improvements of our soil with the increased production. In 1955, a new car, tractor, and combine were purchased. The one-ton truck, in the early 1950s, became a common mode of transportation and work tool for most farm families. It was our only vehicle from 1950–1955 for our family of four, as well as for hauling cattle and grain. It is hard to believe our whole family and a floor gear shift fit in the narrow cab. And that was before seat belts.

The seasons were important for me, for many reasons. Spring was marked with frozen ponds for skating and new clothes and footwear arriving by mail from the Eaton's catalogue. Summer meant school holidays and visits with our grandparents, aunties, uncles, and cousins from Ontario, along with visiting Dad's uncle (Dr.) Callie and Aunt Edith from Brit, Iowa. Harvest, a wonderful time for reaping the rewards of our years' work, was the start of another school year. Finally, winter brought ice cream and more personal time for me.

My farm work, besides the outside house jobs, such as stacking firewood on the back porch, carrying water and snow, helping in the garden, and picking Saskatoon berries for Mom to preserve for winter eating, evolved with my age and abilities. Dad was very particular about building good barbed-wire cattle fences, which meant many hours helping peel off all the bark from the six-inch poplar posts that had been cut to length and sharpened by the large wood. Many farmers only peeled the lower half of the post, but not Dad. Once totally debarked, they were soaked in barrels filled with water and bluestone meant for extending the life of the fence post. Once in the ground, they were placed sixteen inches apart to hold the three strands of barbed-wire fence. A few years later, during the winter months, I chopped down poplar trees along road allowances, thus preventing snow build-up along the roads. I was paid $1.00 per hour by the Rural Municipality of Longlaketon.

I dreaded Easter holiday, even though it was a welcome school break and such a special holiday, because I had to hand pick field rocks before seeding to prevent equipment damage. The size of rock I picked was based on my size and age; I was careful not to injure my back. Rocks were picked and loaded on a wagon. Large ones rolled up a plank and onto the wagon by Dad, which was being pulled by our team of horses, and in later years by our tractor. The rocks were unloaded along fence lines by hand, and rock piles were made on unusable land, such as sloughs. Spring was usually muddy and wet, with our hands getting terribly grimy. Weaving around the field, walking beside or riding on the steel wheel wagon, seemed like a meaningless task. Also, it did not help that Dad had no watch to determine time, determining lunch breaks and finishing the day before heading out along the three miles home by where the sun sat in the sky.

I am sure it is no coincidence that the Anderson mechanical rock picker was invented in Southey, Saskatchewan, fifteen miles from our farm, and made common in the late 1950s by Pete Anderson, who went on to manufacture other modern farm equipment. A tractor was used to pull the two-wheeled machine directly behind it, where a steel grate scraped the ground and, with the help of ground driven reels, bat the rock into a bucket that then self-dumped at a rock pile. Too bad for me it was not invented ten years earlier!

In the early 1950s, we would cut our green oats in July, before the plant stems turned from green to a golden colour, for feed was done with binder making sheaves that needed stooking. Once I could drive the Case DC4 tractor, which pulled the binder, Dad sat on the steel seat to pull the leavers, operating the binder. Our old binder, unlike the newer ones, was run by the tractor and ground driven, and made to be pulled with four to six horses.

One day that summer, Dad sent me out to the east field across the road from our yard to oversee a wiry German fellow named Helmut he had hired to stook oats. I was not strong enough to lift the bundles of sheaves by myself, but I knew the precise way Dad wanted them to lean upright against each other to dry. Helmut soon figured out the exact balance required. His sheaves stood upright in groups of eight; not one collapsed. Stooking is a real art. They were placed in neat, straight lines across the field, making them easier for the trashing crew to remove later.

Finally, it was suppertime. We made short work of heaping plates of Mom's meatloaf and mashed potatoes. Helmut told us, in broken English and using his hands, that he had escaped from East Germany by hiding inside a full oil-tanker rail car. He said he stood up for the duration of the trip, submerged to his neck in the dark liquid before finally crawling out of the container, starving and exhausted, when the coast was clear. Helmut had refused to be part of post-war communism. I did not understand at the time, but Dad said that even in western Canada, Nazi anti-Semitism was real during the Second World War. But he appreciated Helmut's help with the haying and said that he could stay with us if he liked.

Harvest was fun. At a young age, I got to move our International truck around our fields to the Massey-Harris 21 combine, unloading its forty-five-bushel load of grain into our one hundred-bushel capacity International truck. There was a half hour rest period waiting for the second hopper before leaving for the yard to unload a full load of grain into the storage bin. Mom did most of the grain hauling and unloading, which she loved and did very well. Dad ate while the combine was unloading, standing by the open truck passenger door, eating from a metal pie plate, and drinking from a quart sealer Mom had prepared. I remember those long nights when Donna and I were too young to stay at home alone, so we would go riding and sleeping in the truck, waiting for Dad to stop combining. We would go home with the last load, ending another good harvest day. Mom continued to do some of the trucking well into the 1980s. She was known for her lead foot, driving fast and occasionally leaving a pile of grain in the field behind the truck because she had left the back grain shut open while the combine driver unloaded the grain into the truck.

The grain was marketed through the local grain elevator system located next to the rail lines. Dad delivered grain, one hundred bushels per load, with our one-ton International truck to Earl Grey, Gibbs, and Craven.

Research into the history of the prairie elevator system brings up names such as George Stephen and Sir William Van Horne, who worked with John A. Macdonald in building the CPR across the prairies. Grain merchants were granted licences to build and operate standard wood structures along the CPR rail line for shipping farmers' grain to market. Farmers brought the grain by wagon or truck, and the vehicle was weighed. The grain was then loaded into the elevator by a cable hoist that lifted the front of the wagon or truck for the grain to pour out through a back-end gate, following which the vehicle was weighed again. Grain was poured into a pit to be elevated and stored, later to be poured through a pipe into a railroad car and transported east for sale. In the late 1950s and 1960s, I saw a lot of financial suffering from the effects of government control on grain commodities under the Canadian Wheat Board, with trucks lined up for hours to deliver a load of grain valued at $400.00 and needed for family necessities and farm expenses. I remember riding with Dad, lined up on the road out of Gibbs. We were number fifty-one in the line to unload, which took about three to four hours. The Canadian Wheat Board virtually controlled all western grain marketing from early World War II until 2012, when open marketing for all grains came to the prairie farmers. There were heated debates over the years on western grain marketing by the government and their ability to obtain markets that benefited western farmers. The young farmers of today will never know of interim grain payments because of the final settlement that came months after final revenue and cost adjustments were made by the Canadian Government.

—

"Great talent demands great will power," Mrs. Delores Pinkney chalked on the board one day in June 1956. She turned to view our confused ten-year-old faces. "Let me demonstrate," she said. She asked one of my classmates, Gail, to stand in the hallway for a moment, closing the door after her. After leaving her standing in the dark corridor for a brief five seconds, she then threw open the door. "Imagine that my opening this door is a unique opportunity for you," she said. "Now, what do you need to do to accept this lucky, one-of-a-kind chance? Gail looked up at Mrs. Pinkney for a few seconds, then she scanned the rectangular Union Jack—our flag tacked to the drywall above the chalkboard. She took in the silver-framed photo of the new monarch, Queen Elizabeth II, coroneted in 1953. King George's portrait still had a place next to the Lord's Prayer, alongside a posted list of class rules.

"I opened the door for you, Gail. Here, let me open it even more. Now, what do you do when a door of opportunity opens for you?"

Gail's shoulders slumped as she glanced down the hallway to the exit door, the one accessing sunny wheat fields and the baseball diamond outside. Looking again at the class rules, she sighed, re-entered the classroom, walked to her row, and sat down, defeated.

"Exactly, my girl!" exclaimed Mrs. Pinkney. "Fine work, Gail!" she said to the astonished child. "When opportunity opens its portal to you, you do exactly that: you walk through!"

Gail looked around, thrilled at the surprise praise. Her mouth fell open. She laughed.

"Let me also add, children, don't worry so much about what you learn or don't learn here at school." She waved her arm all around the confined classroom. "It's the brave risks you take that open up your life," she said. "Don't worry about collecting the paper stars I give you for memorizing facts. Just work hard to make your dreams come true, whatever they are!"

—

"Grab this weight, Doug," said nurse Smith Martin. "Now lift," I heard as she settled an atomized air cup over my nose and mouth. "Fine. Now raise your right arm again and balance the weight," she instructed, her voice farther and farther away until overtaken by Ron Fiessel's shouts from our old barn.

"Grab the end of this muddy trough and lift, man! This cattle crib is our new boat! We can outrig this thing, Doug," he offered as we figured a way to balance our make-believe boat in the spring slough, such a beautiful tree-filled environment after the cold winter. A couple of times, in the 1950s, there would be a dramatic seasonal change. Imagine going to bed on a cold, snowy winter's night and waking up to the low spots filled with freshly melted snow, complements of a warm Chinook wind.

"Okay, now hitch your horse. Prepare for battle!" Ron yelled. "And bring food, Doug! I'll get the fire going!" Soon, the two of us sat eating our sandwiches in concentrated silence as we awaited the next profound thought.

"Get on the toboggan," I ordered. "Hang onto the long rope. I'll whip you across this frozen territory faster than you ever dreamed," I promised my friend, trying to throw my toboggan rider as I raced through long winter-yellowed grass.

"Stop!" hollered Ron. "Look yonder," he yelled, pointing to our high Ontario-designed barn with its huge flat roof. "Exploration! Look at that air vent, Doug, with the top lightening rod and copper horse emblem. The loft must be fifty feet high!" said Ron. "I'll shimmy up that centre rope to scare the birds!"

Ron, being a bit older and more daring than me, had a hard time convincing me to shimmy up the one-inch thick old weaved rope as he spiraled into the four inch by four-inch air vent that the birds loved. Time with my good friend was a rare treat. A jovial member of the Fiessel family, and only four miles down the road, he was kept busy milking cows and feeding their animals. Two miles from Earl Grey, this wonderful family separated and shipped cream as part of their mixed farming operation.

Our barn

Ron, my fellow adventurer, hurried home to gather the cows from the field using a little red International tractor—something Dad would never have allowed, even if it did not use much gas. I really enjoyed riding with him for the regular milking whenever I would get a chance. I'd watch the separating of cream from the milk with a manual hand-driven machine called a centrifugal separator, after which we'd eat a quick supper before bedding the animals.

Ron and his family taught me a lot. "A gear mechanism causes the separator bowl to spin at thousands of revolutions per minute, Doug," Ron said, "which pushes the lighter cream outward against the walls of the separator. The cream, being lighter, collects in the middle. The cream and milk then flow out of separate spouts into clean milk pails. See?"

"Neat!" I said, impressed. I found out years later my grandpa Woodcock worked at the De Laval factory at Peterborough, Ontario, manufacturing cream separators.

"Dad told me that before hand-cranked separators, the milk sat in a container until the cream floated to the top on its own and then you skimmed it off by hand," said Ron.

"Along with the bugs and dust, if it is not strained through a cheesecloth," Ron's beautiful older sister, Anita, the meticulous organizer of the clean-up crew, added with a laugh. "We do morning and night clean up, Doug," she explained. "Make no mistake: milk can spoil or sour in a heartbeat."

Why go to so much trouble? I thought. We did not do it. To earn a living, even in rural Saskatchewan, was dependent on the amount, and type of, land base each farmer had, along with his interest and expertise, and help that he had available.

"Now c'mon in," said Ron's younger sister, who had invited me for supper and an evening of pre-television family fun and friendship. Balancing a fork on the head of a needle stuck in a bar of soap was a fantastic feat, absorbing a full hour of attention. Another grand treat was making homemade ice-cream in the winter, the likes of which I did taste once many years later when I made it in the same precise way with my grandchildren Jessica, Callie, and Luke.

No matter the dark sky, Ron was up for nighttime fun. "Let us go chip the ice at the dugout, Doug. This vanilla ice-cream will be worth the effort!"

The beauty of stark navy-blue moonlit nights were magical, the two of us with the farm axe, chopping ice under the moon of a pitch-black night. An after-supper card game was standard fun. The adults would be seated around the table, jovial and appreciative of each other, with a teapot and pie both steaming and at the ready for second servings. When it was time to go home, the team was harnessed to the sleigh by the men, brought right up to the Fiessel's door, and I'd tumble in for the four-mile ride home under the stars. Even on a cold night, no matter how blistering the attack of the severe Saskatchewan freeze was in the dangerous pitch dark, the sleigh was prepared with straw and blankets, and we'd make our way along the sled path across field shortcuts to home. On some of these cold winter nights, the northern lights were spectacular, dancing in the sky. Back at their house, the horses unhooked, unharnessed, and fed in their stalls for the night, it seemed no trouble—just another day on the farm. But back at our house, after day trips with our family, the afternoon chores were still waiting and had to be completed in the dark before heading to our beds.

Another winter night, our family and our other neighbours, the Butz family, took a five-mile sleigh ride to visit the Wagner family. The Butz family had four children, and it was a very dark night. The wind had blown all day, so the sleigh path was covered with snow drifts. The narrow sleigh box, with four adults and six children, was too top heavy and tipped over as the horses pulled us over the edge of a large snow drift. We were all tossed into the snow, but thankfully no one hurt and the horse team was well-trained, so they had not run away. The sleigh was righted, and we continued. We had a fun evening and made the return trip around midnight. Jack and Edith Wagner's family were great friends. Although, two of their daughters Val and Gloria, tried to keep in touch, I was always too busy to write back and have very much missed and regretted supporting our friendship.

I was lucky enough to be able to join the Boy Scouts in Earl Grey with my friend Ron and several boys from town. Mr. Bill Teece, the railway station master, was our leader. Our meetings were held in the train station waiting room where the centre pot-belly woodstove sat. Next to the station master's office, we could hear the teletype clicking. The passenger train had just ended with only freight services provided but still by the coal powered steam locomotive. My generous eastern relatives supplied most of my uniform and equipment; they were so good to us. For the next few years I was involved, we did projects like kite designing, building, and flying, along with learning survival exercises. Being out in nature, which I loved, and learning survival skills were important. We actually built small three-inch-deep dugout huts, with weaved straw roofs from the combined wheat straw on the Fiessel's farm, and spent a night sleeping in them to understand the environment of the early pioneers.

One winter evening, after school, I tried to ride my pony to Earl Grey for our monthly meeting. A storm was quickly arriving. In the dark, following snow-covered slay paths in strong northwest winds was impossible. These slay trails wound through neighbour fields and not in any straight lines. Soon, with no clear trail and the huge snow drifts, my horse and I had no alternative but to return home.

—

"If you'll shift a little to your right, Doug," spoke a caring voice, "let us see what's going on with the patchy redness on your shoulders and back."

I launch into an easy explanation, even though I sense my mouth is not moving. Am I asleep? In any case, I explain that farm boys like me quickly learn to distinguish between the spring and hot summer sun and winds. I measure the toll of the sun on my burned skin and ears not by the degree of burn, but by the depth of needle-hot pain. Excuse me if I sound a little specialized here, but there was no sunscreen or concern over the effects of the sun on our skin causing problems such as cancer. The concern was the heat affecting our working ability. Therefore, if possible, we were to wear a wide brim hat or some sort of cover to shade us from the sun.

I watched the harvesting of cereal grains change in the 1950's from straight combining with a twelve-inch header, to swathing, which is laying crops in a swath or strips and having a pickup header on the combine. Our first swather was sixteen inches pulled and powered by our tractor, which sped up our harvest considerably. The idea of swathing and leaving the crop laying on the stubble to a final dry, rather than standing before combining, was nothing less than revolutionary.

"Radical," said Dad, his brow wrinkled in doubt.

"Visionary," I replied.

Dad purchased his first used pickup header from the Regina plains, where the swathing idea had taken hold a few years earlier. Our Massey-Harris 21 self-propelled combine had a hand lift (wheel) to the right of the operator, with a foot brake. I learned the operation in no time by grabbing the wheel on the right, pushing the brake off with my right foot, and either thrusting forward to let the table down or pulling for the table to raise. The lift weight, thus easing the hand turning of the wheel, was made easier by three large springs that could be tightened or loosened by a long bolt with a lot of threads. The tightening, though, could be done only by laying down on the sharp grain stubble underneath the combine, causing sore red patches of skin all over my back. There was no need to move the springs without the pickup, but with the extra weight of the heavier pickup every time it was installed, the springs had to be adjusted by us. To do this, we had to be on our backs, on the ground, in a very enclosed space, wheedling a large wrench and turning the large rod nut. This was scary for me, but I became good at it, and the experience held me in good stead for the future. Much later in life, I would wield weighty trailer doors of unusual sizes with all the patience required to get the job done. During the first few years with this new machine, only part of our crop was swathed, so the demanding process was repeated a couple of times a year.

—

Our family faith centred around the Earl Grey United Church that Grandpa Maurice attended on arriving in Canada. Sister Donna and I attended Sunday school and later sang in the church choir while Mom taught Sunday school. Aunt Grace, Mom's older sister, was a missionary and teacher. She would visit us on her furloughs from the leper colonies in Bolivia, South America. A little extra religious teaching for us and our neighbourhood children happened with summer camps on our front porch, although her loving blessings at mealtime were usually dropped after she left, except on special religious days. At our United Church in Earl Grey, Aunt Grace did not spare hardworking farmers of her daily grind in the leper colonies and its tragedies. Hopefully, these talks sent complainers home with a lighter load. Aunt Grace had a profound effect on my thoughts around the concept that everything was possible. At Mom's insistence, Donna and I joined a fun young-peoples group and were warmly welcomed by the leaders, Ailene Tallentire and her husband, Bob, manager of our local Co-op Grocery store.

Grandpa Brewster's voice boomed at our Sunday United Church services during my Sunday visits with him. He was good to me, and I missed him terribly after he passed away in March 1958, when I was 14 years old. Just before his funeral service began, I found myself walking alone into the church where my kind-hearted grandpa laid at peace. My seeking him out and seeing his black eye patch and his large, immobile body profoundly affected me. My hands sweat, as astounded, I was bushwhacked with the reality of human extinction. There we were, both with bodies, but one of us no longer pumped blood. The notion of time's perimeter hit me like a ton of bricks. A lifetime has fences, too.

—

"Good evening, Mr. Brewster," said a cheerful hospital volunteer just as I returned from my physio routine. "Would a Saturday movie be fun? Maybe some family time?"

"Saturday's fine! The first TV in town," I hastened to explain, "is at the Co-op store. Yes, indeed, it was common for forty adults and children, mostly town locals, to sit around one TV until well after closing time. The store manager, Bob Tallentire, distributed bowls of popcorn row by happy row. Until then, the town hall was our only means of seeing rented-reel movies, just as our six-volt radio, tucked in the corner cabinet at home, provided our sole source for news, weather, mystery stories and sports."

Soon, Grandpa and Grandma had their own TV. Our winter trips to Earl Grey for supplies and a visit introduced us to the first modern home television. Saturday night shows like"The Ed Sullivan Show," "Alfred Hitchcock," and "I Love Lucy," along with Roy and Dale Rogers, Gene Autry, and the Lone Ranger and Tonto, were all the modern cowboys living and producing movies in California in the 1950s, which I was lucky enough to see in Earl Grey, Saskatchewan. By 11:00 p.m., Dad would bring the sleigh and horses from the school barn next to Grandpa's, and we would load up and head home for the long seven-mile ride. It was an enjoyable day. We had our supplies: a large four-gallon vanilla ice cream treat. If we were lucky, it would keep frozen in our porch. And we would be home soon with supplies unloaded at our house and horses and other livestock taken care of before a good night's sleep.

—

The spring of 1957 again was another reality check for my sister and me. We had been a family of four with not a thought of anything else at least for sister Donna and me, when Mom asked, "What would you think of a new baby sister or brother?" Well, after the shock, and seeing Mom so happy we soon embraced the idea of us having a new member to our family.

On October 5, 1957 Darcy Dwight Brewster was born, with everyone healthy and well. Darcy arrived home turning the family's routine up-side down. Being a teenager, I am sure I probably ignored him for the first year, however sister Donna soon bonded with her baby brother and was a great help for Mom. Although, I left for University a couple years later I returned to work on the farm during the summer break until 1964. Looking back, even though we spent time together, I expect it was limited, with me busy working and spending time with friends. Knowing Dad enjoyed outdoor wiener roasts, I would arrange for young brother Darcy and me to gather wood, start a fire, and cut live willow wiener sticks for a family wiener roast, it was great fun.

At fourteen years old, the same age Dad was when the Great Depression darkened his youth, my life was brightened by the installation of electricity in our district. I had no idea that Saskatchewan's Co-operative Commonwealth Federation (CCF), a socialist political party in Canada with our publicly-owned Saskatchewan power corporation, mandated the free electrical installation to all rural farms, a convenience not expected until the prosperity of the 1950s. Soon, there were tall grey power poles carrying the stretched electrical wire from one farmyard to another, through farm fields and yards. Little did we know that someday these poles would become a real problem for the future of agriculture. Grain farming, especially with its much larger equipment, has a problem maneuvering around them, which causes safety and damage issues.

The yard buildings were wired and prepared prior to the installation of the transmission lines. The farm houses were usually the first buildings prepared, along with barns and shops. This part of the electrical installation was paid by the individual farmer, so not all barns and outbuildings got wired.

When the electricians preparing the farmhouse for electrical service arrived, Dad was very obliging and helpful. He was always interested in new things. But the installation was no easy task. They had to find a location for the central fuse box and distribute wires for light sockets, switches, and receptacles, which service various appliances. The wires had to be run through finished walls and ceilings, hidden from sight and protected from potential damage. Crawling into old attics and crawl spaces meant surprise confrontations with bats and other rodents, those spaces being their favourite hideouts.

Insulation, if there was any, comprised of old newspapers and wood chips. I was fascinated by seeing behind the veneer of plaster and the thin lath boards with the various cubby holes that had been next to me while growing up. Every spring, along with changing outside screens or storm windows, Mom would have to clean the paint and wall-papered walls and ceilings, due to smoke and soot from burning wood and coal.

Dad, being quite involved in the process, educated himself so that he could wire our big Ontario-style barn himself. A farmer was allowed to obtain a permit from the government to wire his own building, so Dad obtained a permit and printed manual detailing the installation. With his hand tools and tips from house crew, he wired our large flat-roofed barn. Apparently, his work passed the electrical inspection with flying colors!

The electrical crew returned with their heavy equipment that fall to install a yard pole and hang transformers for the final hook-up. Before turning on the power and connecting the house to the transformer, a yard light was installed at the main pole. Working in a field near the yard, monitoring the progress, I finally drove Dad's seventy diesel John Deere tractor and fourteen-foot John Deere cultivator into the yard, excited for the "let there be light" moment! I knew about the lights in town and six-volt farm lighting systems, as Uncle Calvin and Aunt Jean had them at the time. I listened and watched with great anticipation as the switch turned on our electricity. Soon, the sun would be gone, and darkness would be upon us. I tried to imagine what our house and yard would be like with our new electrical power. I had experienced electrical lighting, but this was different, because it was in rural Saskatchewan. I had only experienced darkness here, from a young age, with dim wick or mantel house lamps and hanging chore lanterns. I wondered, because I was never comfortable in the dark if it would give me added security and confidence. My wondrous look at the Electrician, I guess, made him recall his grandpa telling him about the first time he heard a real human voice coming through a little handheld receiver attached to a wooden box on the wall—a magnetic six-volt battery phone. We still had it hanging on our kitchen wall, which meant we had a phone line connecting us to the neighbours and a central telephone exchange.

"Pretty soon," he told me, "these here gas lanterns and coal oil lights will be antiques." I think he liked me, because as we both eyed the geese settling down across the road on a grain patch, his smile told me he was all for advancement and new things.

Who knows, I thought, *maybe someday I can help improve our lives, too, and make things easier for us.*

—

"Hello brightness!" said Mom, serving the crew chilled lemonade and the flakiest rhubarb pie ever. "Sure, a single functioning oil lamp kept our family close together," she said, "but it's wonderful for my children to extend their daytime hours."

The TV was just the beginning. Mom's brand-new electric stove replaced her wood burning one, and an electric washing machine made her life so much easier. One day after school I found Mom sitting in a straight back chair, smiling at the old green gas-motor washing tub as though settling in for a concert performance. Goodbye to its exhaust hose, which wound its way outside on washing day because of toxic fumes from the running motor. The gasoline storage container was so dangerous. Mom gazed reverently at the electric single rotor and wringer that would make doing laundry so much easier. However, unlike the modern washing machine, we still had to heat buckets of water or snow, reuse the soapy water for a few batches, and drain the dirty water back into a pail to lug outside.

When the gravity flow oil burner replaced our wood burning central heater a couple of years before electricity, Mom was happy not to have more wood to contend with in our living room. When electricity came, Mom welcomed the next miracle one fine Saturday morning, which was the central flow furnace with an air-blowing fan and fuel tank outside of the living room. Still, while thrilled at the advantages of electricity, when a power outage occurred (and several did), Dad reminded us that we farm folk had our fully functioning antique options. Having ousted the wood and oil burning stove from its mounted place of glory in the centre of our living room, and relieved of the continuous cleanup and carrying of wood and oil fuel, our new oil-burning furnace fanned out radiant heat to all the rooms.

"Hello modernity," sang Mom.

—

"Doug! I'd love to see you look as excited about school," Mom said.

"This is school, Mom," I replied, skillfully revving up and reversing my tractor, heading back to the field.

"Lights in every room of the house?" asked Donna, quickly assessing the advantages of a little privacy. "You mean, I can read in bed and get dressed with a light on?" Donna loved to read victory stories of overseas service organizations, such as the Red Cross. This passion may have been inspired by the caring work of Aunt Grace in Bolivia, or by the volunteer youth groups organized by the Tallentire's at the United Church. Donna's lamplight pooled a yellow semi-circle into the hallway many a late night. I made a bed headboard reading light out of a large syrup tin by cutting it in half, soldering metal hangers to it, and attaching a light socket and bulb. I hung it on my bed for reading.

Electricity gave farmers tremendous gifts and conveniences, such as heating elements to thaw frozen motors in winter weather, power tools, and welders for outdoor repair. Up until that time, Dad did his building with two hand saws: one for ripping along the grain of the lumber, the other to crosscut, level, hammer, and square. I still marvel at his craftsmanship. He took such pride in caring for even his most basic tools. I learned on our first welder, which operated by belts off the pulley mounted on our Case DC4 tractor. All holes were made with a variety of hand and wall-mounted drills, with larger holes in wood being burned through with a red-hot rod.

I would be remiss if I failed to talk about the other personal necessities needed to stay healthy and survive the surrounding elements. Dad hung on to the old ways of doing things, thus we were usually last to get new and improved personal conveniences. Our old kerosene lamps were used until the installation of power in 1957, when many famers, such as Uncle Calvin and Aunt Jean, had six-volt battery systems with windmill charging systems. The house had its first improvements in the early 1960s, so even with power, we did not have a bathroom with a flush toilet or bathtub. A large, round galvanized tub had been common for many rural families, but in our house, adults washed themselves with a basin of water, soap, and wash cloth. The toilet of the time was an outdoor building over a hole in the ground, with a wood bench, an opening to sit on, old catalogues for paper, and of course, a door. A small, basic indoor storage container evolved from a chamber pot under the bed to a seated five-gallon potty. But again, I would expect our daily walks to the outdoor biffy continued longer than most rural families.

—

The principal of the new Earl Grey School was Mr. Gordon Size. The vice-principal was Mr. Code. Both were fine men. In my last year at the school, Mr. Elson, an older gentleman who had been through the trenches during both wars, helped me to finish up my final year of high school.

These school years, especially from 1958 to 1961, gave me more access to my community. I had been quite shy before this time, but during this time I began to enjoy the outside world. In 1958, three of my friends (Jim, Gordon, and Barry) and I purchased black leather motorcycle jackets with big bald eagles embroidered on the back. In junior high school, we tried smoking in and out of school, sometimes trying dead leaves. On Saturday nights we would find a meeting place in one of the old vacant business buildings, sometimes even with power. It was great fun, and the odd time we even purchased a large bottle of vanilla! I often wondered what the local store owner thought as he sold it to us.

Our intent was never to damage these places, and in most cases, we left them as we found them. We were not part of the larger slightly older "in group" of more worldly guys, especially when it came to flirtation. I expect that was one of the reasons our group formed. During my last two years of high school, I was friendly with everyone and joined school activities, such as curling, football, and track and field. I even played in the Saskatchewan Provincial Curling Championships one year.

Mr. Elson's terrified students failed to notice me high atop this pole at Vic and Rosemary's farm, but I would see them running around like scared rabbits below. First, Miss E. Swanson, a Fosterdale teacher, would whip by in the ugliest car ever made, a Studebaker. Even with my excellent vision, it was hard to tell the front from the back. The car was all windows and vomit coloured. But back to Mr. Elson. He served in World War I and then, amazingly, he fought a decade and a bit later for the Brits in World War II. Grandpa says Mr. Elson smoothed the way for a lot of kids to graduate. He knew where the real classrooms were, but man oh man, he would tolerate no disrespect, especially destruction.

"The next morning Principal Elson threatened all the boys at school with a visit to Judge Elliott in Regina for a round of lashes for their part in the disaster I see below. Bales of hay strewn in the house, unfenced cattle, every pail kicked aside, porch chairs upside down and tossed off the wrap-around balcony and into the gravel... amazing how rapid the clean-up is. I'm just ready to slide on down here to a dozen teenagers sweating silver bullets.

I do not see that Brewster kid. I'll wager he was too busy doing chores at his family's farm on Halloween night" said Mr. Elson.

I think Mom and Dad also helped more than I appreciated. Dad had always talked about his Grade 10 year and how important education was. When I was in Grade 10, Dad and I went to the offices of John Deere in Regina to ask about a job. They advised that I stay in school, which was significant for my future.

I studied accounting by correspondence, enjoyed it, and got great marks. Grades 9 and 10 were a struggle for me, and without the excellent support I received, I would not have remained in school. Towards the end of Grade 10, I began to think about the future. Would I farm or get a job? Dad had already crossed off farming from the potential occupations list for me, but he never came out and said, "I don't want you to farm."

On my Grade 11 exams, I was above the middle of the class and close to an eighty percent average. I had also been trying to catch up on French after school, as it was a requirement for university. Mr. Code, the French teacher, would stay after school to help me. I most likely still had a little too much fun time over schoolwork, but I honestly think I needed that for my personal development. I signed up for summer school in Regina for my Grade 12 French. Those six weeks were long and hard, but I succeeded with a passing mark.

The Grade 12 marks finally came out in July, and I had passed with marks to qualify for university. Life was great! For the next three summers, I was around the family farm for most of the livestock haying, baling, and swathing before having to leave in the early fall for university in Saskatoon. I felt guilty about leaving during such a busy time, but I wanted more.

The farm required my help, as it was growing and expanding. When I did decide to leave, perhaps Dad was disappointed, but he never tried to sway my opinion. I never knew his true feelings. Dad was always hard to read, while Mom was strongly supportive in whatever I did. I believe Dad most likely was too.

—

"Dad look up! It's Mr. Woods!" Working in our farmyard that last summer before leaving for university in Saskatoon, Dad and I waved to the ambitious fellow who, only a few years earlier, earned his teaching certificate in Saskatoon. He was high above us, piloting his small plane in large arcs toward Fosterdale School where he had taught me in Grade 5. He helped to build a power dam now called the Gardner Dam.

"Radio says they're aiming to call the huge water basin Lake Diefenbaker." said Dad, leaning on our truck.

"Ken! Our electricity flickered off," Mom said with a wave from the kitchen. "My washer's off. The stove, too!" Hours later, with no explanation, the washer resumed its spin and the stove light beamed red and finished baking the collapsed bread in the oven. Only the next morning, at the beginning of her waitressing shift at Elsie's diner, did Mom learn the tragic cause of that outage. Mr. Woods, unfamiliar with our new electrical wires, flew too low. His wing caught a cable hanging only three hundred feet in the air over the school where he'd taught so well. He was killed instantly. A crew collected the strewn damage, the broken wing, and the pilot's console, but my grief scattered afar. How fragile is this life? A switch turned on and off. A temporary beam. With every new positive, there is a negative. With light, there is darkness, too.

UNIVERSITY AND BACHELOR OF COMMERCE

> I come exuberant,
> With boundless youth,
> Expectant
> Of what lies ahead,
> But eager
> To experience my new surroundings.
> Excited,
> I pass through the gates
> To the University of Saskatchewan
> 1961 – 62 (The Graystone—U of S)

"THERE'S NO REAL CHOICE ABOUT GOING HOME to help Dad with harvest, Jerry," I said as we teens hauled our gear up from his Dad's car to the second floor of our boarding house bedroom. "It's 150 miles to the farm. It's impossible to bus it, work all weekend, and be back here on time for my classes."

"Look at this place, Doug. We are among a dozen students in this huge house. We have been here ten minutes, and we each have a key and a printed list of house rules. We know where the washer is, and supper will be ready at 6:00 p.m."

"Smells good, too," I replied.

"Looks like a hearse in the driveway," said Jerry, drawing back the lace curtain. "I hear there's a tour in it tomorrow. Maybe we'll drive over the South Saskatchewan River flowing all the way from the Rocky Mountains."

It is the end of August 1961. Here I am in Saskatoon for my first year of university. I am eighteen years old.

"Just fifteen blocks to school," said Jerry. "Let's jog over and back before supper."

"Eleven o'clock," whispered Jerry. I glance up to my right. "Three losers looking to bully the newbies," said Jerry.

"Wanna test those wings on yer fancy leather?" sneered one guy, referencing the bald eagle emblazoned on the back of my jacket.

"Eh, farm boys?" called out another from the wheel of his poorly parked jalopy.

"He wouldn't have the first inkling of how to milk a cow or drive a tractor," I commented to Jerry, but we did not engage. We did not make eye contact. We covered the fifteen blocks to school, hit a used clothing store, got this country boy a warm city coat, and jogged home on time for a delicious supper.

Get set for change, country boy, I advised myself.

The boarding house owners had it down to a mini science: four bedrooms, three guys to each room, tasty meals, laundry, mail, keys, and even an ancient hearse we'd all pile into for an occasional outing. Outside of a commerce classroom, I learned about running a smart business. I retired my motorcycle jacket that night.

The College of Commerce operated out of an old air force hanger on the campus, located some distance from the main campus, which featured the beautiful stone buildings of the University of Saskatchewan. Soon I would be running from one building to another, participating in classes. On registration day, all freshman commerce students lined up for orientation to choose our courses and pay tuition. Wayne Quinn, the local Co-op's manager's son from Earl Grey, was a senior student finishing up his last year of his Bachelor of Commerce degree. He noticed my tuition cheque did not have the amount line filled out properly and explained that one should always draw a line before and after the written amount, to ensure no changes can be made to the cheque. It is a practice I never forgot and passed on to other people. I shook hands with another past Earl Grey student, Gilbert Wilde, whose Dad, Adam, renovated our house in the 1960s. Gilbert had also studied accounting and business in Saskatoon.

The first week of introducing first year students to University life was great fun and very informative. There were campus tours, college parties ending with a city-wide interruption lasting for hours. The students formed a human chain through downtown Saskatoon, holding hands to form five miles of unbroken resolve through buildings, streets, and city buses. I learned a social lesson about the power of human connection and determination. The more who joined this human rope, the further the disruption, which was a profound message of political activism. While I went to Husky football games and drama productions, and listened to speakers such as John Diefenbaker, my activities outside of school were limited due to my primary goal of passing my courses.

People from many nationalities and walks of life were visible on campus and in our classes—both professors and students alike. Degrees of intelligence, class levels, attitudes, wealth, and up-bringing melted in the singular determination to do well. A sense of competition surfaced. I learned so much about different political parties, and I began to consider where I stood on important issues. I began to walk in a whole new world.

With first-year exams over, I was on my way home to the farm, rolling along in the Grey Hound Bus one wintery April morning, scene after scene passing by the window. The bus leaving Saskatoon passed the boarding houses where I had lived, including the one with the chronic noisemakers, which prompted my decision to move to a quieter place. We drove past my favourite spot to sit and watch the ice breaking, right alongside the rolling Saskatchewan River. Through her kitchen window, I could see my land lady's bent shape as she prepared the evening meal. I could see the old lumpy pull-out couch placed out on the curb where I used to wait. I wondered if that old card table deck was also being tossed. Snug in the bus, I recalled my strolls to the river's edge where, in the warmer weather, I would fold my heavy winter coat to sit and track the fast-flowing leaves, metal, or feathers bubbling down the river or stuck on a piece of wood or ice. The chunks got carried away by the river and yet, despite choppy waves, even the tiniest leaves clung on for the long haul, sometimes by the slenderest of spider web filaments.

The bus slowed down, followed by the expiration of air, a prolonged sigh like the deflation of an air mattress. Decompressed doors opened. I was blasted with cold air. "Goodnight, now," the driver said, dropping me on the side of Highway #11. I lugged my suitcases, boxes, and university satchel across to Highway #20 where I could hitch a ride home.

It was late in the evening, dark, and getting colder. I needed to act, so I hid my gear-including the used typewriter I bought that past winter, despite Dad's disapproval, in the two-foot ditch alongside the highway. With the lighter load, I continued to trek home, hoping for a ride from each passing car. Two exhausting miles up the road, the faint outlines of the Wolf house appeared, honey yellow light spilling out of their window and onto the wet spring ground. Unfamiliar with their yard and anxious about the temper of their family dog, I cautiously approached the front door. When I was ten and selling garden seeds door to door in Earl Grey, I almost got bit by a violent, sharp-toothed, saliva-dripping dog that was stopped at the last minute by the forceful restraint of his owner. I am sure I had backed myself halfway up their front yard Elm tree trunk.

"Doug! Welcome!" said Elsie Wolf, an energetic daughter of the Brandt family, who grew up down the road from us back in the 40s.

"Exams are over, Elsie. I'm heading home for the summer," I said, happy to see her.

Long married to Charley, and cheerful as always, Elsie sat me down to a generous slice of warm pie and a good, strong cup of tea. I downed the lot before a thick pan-fried grilled cheese sandwich slid snuggly onto a plate and under my nose. With the first inhale, I was transported yet again. I could see Mom at the restaurant counter—the high one I could not yet reach. She was helping Elsie and her sister at Lumsden's best eatery. Amidst the clatter of cutlery, I could hear laughter.

—

"There are two reasons I like to work at Elsie's diner, Ken," said Mom. "One: I like the extra cash to fritter as I please without having to explain myself to you. Two: I like friendly company, plain and simple. I love Elsie. I love her sister, Ida," Mom said, noting a slight frown crease Dad's brow. "I love the whole Brandt gang, a genuine set of friends. We have fun together. We treat ourselves to a milkshake or an exotic slice of pizza instead of penny-pinching for a shiny new tractor for the farm. I need to widen my circle, Ken!" Tracking her husband's slow glance toward the laundry basket and trailing it to the empty cookie jar, Mom added a retort that sent shivers down my spine and caused Donna to cry. "And, of course, down the road a bit for a few hours a week, Ken, is a lot fewer miles than Peterborough, Ontario." End of conversation. Mom's threat usually did the job. We knew without a doubt that if Mom up and left, we would be going with her. We'd be saying goodbye to Dad.

—

In Elsie's kitchen tonight, so many years later, I reminisced with my host about the long days of waiting around that Donna and I spent while Mom worked at the diner with Elsie and her sisters. "I remember," said Charley. "Occasionally, Doug, I'd tuck you and Donna into the little travel trailer behind the restaurant for a rest in the afternoon. My mini-home on wheels!"

"Thanks, Charley," I replied, suffused in rosy gratitude in remembering our shared history on this dark spring night in the Qu'Appelle channel. I knew Dad was on his way to pick me up.

"You know, Doug, you're a bright young man," said Charley, smiling broadly. "I noticed you've ditched your fearsome bald eagle leather jacket."

"Not that anyone was terrified," said Elsie with a smile, refilling our cups.

"Have you heard about the social credit political party on campus, Doug?" began Charley as he leaned forward. I knew he'd pulled out a political fishing pole, hoping to reel me in. But when Elsie turned to her husband, the forward lean of her shoulders and her face showing interest, I did the same.

"Our Saskatchewan province formed only fifty-six years ago, in 1905, and it encountered difficult times after the financial crash of 1929. You know that, right, Doug? You know how Ottawa dusted us off with their pretentious praise of what fine God-fearing Christian farmers we are—upright, steady, and stoic through any drought, devastation, or windstorm with which God elects to test our unfaltering faith. I hated their bringing God into that hooey! What I loved, though, is how Premier Tommy Douglas rescued our province during those terribly hard times of the 1930s and 1940s. He was a democratic socialist and became premier in 1944, one year after you were born, Doug. He pulled Saskatchewan out its dire financial ditch.

"Over his tenure there have been many improvements to our quality of life, such as the government electrification program, provincial medical insurance, and many other government assistance programs that work to improve the lives of the working people. I've been studying the social credit philosophy and believe their ideas on government-controlled federal spending have merit."

"Your ride's here," said Elsie as the slow crunch of tires on their wet gravel driveway signalled Dad's arrival in his blue 1950 International One Ton truck. "He's not leaving without a cup a tea." Dad obliged with his usual politeness and quiet appreciation.

The three of them enjoyed a good laugh about the long-ago day Elsie's Mom chose Dad, a cautious driver, to give her a lift to town, insulting her dare-devil son, George, who raced down the local un-graveled roads, pedal to the metal. What fun, then, for George to be called upon to haul Dad out of the rut he careened into because he had followed the snow tracks ahead of him instead of the curve of the road.

"George had the good grace to keep a serious face," Elsie said laughing.

At the door, Charley took me aside for a moment. "Doug, if you'd like to hear more about social credit, you know where the cherry pie is! And remember, your mom named you Douglas for a reason!"

Driving home, I told Dad about Charley's social credit ideas. Dad nodded and mentioned that, in the morning, he would show me his repair of the broken cultivator axle. He said the cows were going to calve soon, and that after the morning chores were done, I could use his truck to get my belongings out of the ditch near Lumsden.

Once inside the front door, I turned to say goodnight to Dad, but he had already retreated to his bedroom. His door was closed.

—

I returned to help dad on the farm and earn money for school during the summer breaks up until 1964, when a rough Saskatchewan windstorm cut through the Earl Grey farm district at a speed that devastated our buildings and crops. I was shocked. The hay rack, grain auger, and buildings had been moved and spread across and out of the yard. In seconds, with no more than groans and creaks, invisible force flattened many trees, shelter belts, and grain crops.

That fall, Garry, one of my roommates, did not return to university because his family needed his help to repair their devastated farmyard, which had lost granaries and the barn roof to the windstorm. He said their flock of one hundred chickens was swept up and disappeared into the sky.

The topography of our agriculture area was rolling with lots of poplar bushes and spring water sloughs. The land clearing we were doing meant cutting, piling, and burning trees. The clearing up of sticks and rocks was done by hand. We had an old Allis-Chalmers HD9 caterpillar for the bulldozing of the standing trees. I finally convinced Dad to buy a loader and a John Deere tractor to help with lifting fallen trees and scattered rocks. Imagine a farm without a front-end loader! Previously on our farm, heavy planks or sturdy poplar poles were the only means to pry, move, and lift a heavy load. For example, at butchering time, Dad stood three poles up together with the bottoms apart and tops chained together with a pulley system for lifting up to fifteen hundred pounds.

To help move the rocks off the cultivated land, I designed and welded together a hoist trailer out of a scrap trailer. It did not look like much, but it dumped the rocks in a pile without manual work. Also, by converting an old small Graham Hammy cultivator into a tree stump rack, they could be bunched in piles. I learned to weld on my own, and although it did not look very pleasing, the welding held the steel together.

—

University life was not only about studying; I also enjoyed getting together with friends. On weekends, especially after first year and me getting my own car, a black 1954 Ford, I loved to drive around and go to outdoor drive-ins, the outdoor A&W and more. From the time I was a teenager, I enjoyed dating girls and going to dances; however, I never made it a serious matter.

Although a bit of a loner, I enjoyed friendship, and I had wheels from the age of sixteen, which was a great benefit. In other words, for a girl sitting at home with her family, getting a call for a ride to some event in rural Saskatchewan would often give her a smile. She could ignore my faults if my vehicle got her to Point A and back home again. I had fun and met some great girls. I would not change anything about those years.

The University of Saskatoon has rows of boarding houses all full of students. In my third year we had a second-floor apartment, and the next house was full of college girls, so it did not take long before we got to know them and were having fun together. I noticed one girl. She was quiet and stayed away from the main group; she had her own apartment. Although not overly friendly, I managed to get her name: Merle White. She was a pharmacy student in her third year from Instow, Saskatchewan. She seemed quite different from the regular collage girls, and I was inquisitive and admired her composure and directness. It was slow going but eventually we became friends and started dating. I was content, admiring her maturity and focus on educational goals.

A Lucky Prairie Boy

In the fall of 1963, I moved closer to the Saskatoon campus and shared a basement suite in a house owned by Doctor Edwards, a surgeon at the university hospital. Every Saturday he did a simple, heartwarming thing. He baked the rolls he and his wife would share with us.

One morning that fall, while munching a warm buttered bun brought downstairs by Dr. Edwards, I signed up for a student loan and paid $20.00 to Frank Flaman, an entrepreneur from Southey, to make the arrangements for me to pick up a new car in Toronto. Much earlier, in 1957, Dad had traded grain to Frank for our electrical appliances. Mr. Flaman later went on to expand and diversify his farm's sales operation, which now covers western Canada. But it was time for my own vehicle. Like my dad, I believe that investing in quality machinery was, and is, money well spent. I did not discuss the disappointment I felt that Grandpa's 1953 Ford sat in his garage, having not moved since his passing. Dad and Grandma had declined my offer to buy it. Later, I was pained to discover that, long after my need for it had passed, Grandpa's car was sold to a local farm friend.

1965

I purchased a train ticket leaving after my Thursday classes, from Saskatoon to Toronto coinciding with the delivery of my car and Mom returning from a holiday with her family in Peterborough. I enjoyed the fantastic scenery through Manitoba and the Great Lakes and Canadian Shield area of Ontario. On arriving late Saturday morning, I met Mom and had a short visit with my Aunt Gwen and Grandma Woodcock in the huge Toronto train station before Mom and I said our good-byes, leaving by taxi to the car dealership. I paid for my new two-door Ford, got the license plate installed, and stood in awe admiring my baby-blue car with lots of chrome.

As I detailed the considerable changes in automobile designs between 1964 and 1965, Mom extracted a bag of pre-packed sandwiches and urged me to hightail it to Saskatchewan.

Behind the windshield of my new purchase, I happily adjusted the radio dial on the clean console to a country music station on the Dominion and Trans-Canada network, recently absorbed by the CBC Radio Network. I backed out of the Toronto dealership's lot and focused on the quiet discussion going on in my mind.

"Just drive, Doug," I told myself, in response to my question about where I was headed. "All things considered; your studies are going ok. You attend your commerce classes," I continued, "because you are not one who can skip class and still manage a good grade. Look, you can graduate this year, but you will need to redo a course. Take the six-week summer intensive and be done with it."

Slowly braking, I detected a faint disc resistance as, turning the wheel, I rounded onto Strathcona Blvd., heading west.

Mom and I left Toronto the same day I had arrived, driving straight through nonstop. From our perch, the blitz ended as I steered into our farmyard in the darkness of the wee morning hours. Dad, Darcy, and Donna were fast asleep. After the long trip and crashing at home for seven hours, I tore off to Saskatoon for my Monday afternoon class.

—

I was so fortunate to study at the University of Saskatchewan campus in Saskatoon. The College of Commerce was established in the early 1940s to address the need for business education in the areas of accounting, finance, marketing, human resources, and management. I missed a new Commerce and Law facility built in 1968, with its modern, up-to-date facilities rather than our old war hanger building, which was prone to floods. Eventually, in 2007, the university's business program and the Edwards School of Business would amalgamate under the name of the latter.

Time moved on and I completed fourth-year exams. Although I passed all my courses and was in the graduating class of 1965, I needed one more subject that I had missed in first year. I had been interested in commercial law, so I decided to enroll in law school that upcoming fall to pick up the class I'd missed and complete my first year of a law degree program.

I continued seeing Merle, who was busy working after classes and

Doug Brewster with His Bachelor of Commerce

on summer holidays at the university's chemistry laboratories. Unlike myself, she had no doubt about her chosen path and was sailing forward with her career plans.

To save money, I moved into a smaller riverside boarding house. I also applied for a trainee summer job at Federated Co-op, a large wholesale company supplying the co-op retail stores throughout western Canada. My respect for Bob Tallentire, who had worked and represented the huge benefits of the rural co-op services, and my keen interest in the co-operative business structure helped me get the job.

At Federated Co-operatives Limited (FCL), I checked the company's expense approvals. At my small corner desk one Monday morning, I reviewed the company's president's expense invoices. Several of his claims missed the proper documentation. I headed up to the top floor corner office and asked the secretary if I could see him. Wide-eyed at my request for the proper documentation, the president floundered a little bit, got me what I needed, and said he was impressed with my initiative. His acknowledgment felt good. Even a single pat on the back, no matter how rare, can encourage self-confidence.

—

"Let's do it, Norm," I said to my friend. That winter, we constructed a sturdy raft to sail down the South Saskatchewan River. This was my first mobile without the underpin of wheels or runners. We planned to intersect at the north tributary near Prince Albert, and somehow our adventure began attracting media attention. We received free forty-five-gallon barrels from Federated Co-ops Ltd and used lumber from Grandma Hazel for the structure. We much appreciated these gifts that June.

Finally, after much hard work and setting up the tent and a fire pit, supporters gathered for our launch. Although worried about the river's undercurrents, we forged ahead confident in our design. After pulling on our weatherproof fishing boots, our canvas vests, and life jackets we gave a toast with our tin coffee mugs and shoved from shore on the raft we had dubbed Jonah. With wood, food, and life jackets, we cast off. The small crowd cheered from behind.

"Wonderful, Norm!" I shouted. The weather was overcast, and the river fast moving. Soon we beached close to a fire pit, but given the pouring rain, we took refuge in our tent. Looking up into a brighter sky a little later, I saw two large-winged birds part company, one soaring to the left and the other to the right. Time and again the raft gravitated toward the underwater sandbars, submerged shoals of stubborn sand. Nature held sway as she taught us about the amount of water a fast-flowing river requires to keep us afloat above its sandbars. After two days of arguing with nature, we admitted defeat.

The end was simple. We ditched the raft, walked to the closest road, and hitched a ride home to Saskatoon. When I think of that defeat, it's always in connection with my marriage, also stalled by hidden forces, such as an unresolved history perhaps. That sandbar blockage prevented any sailing off into the sunset.

—

Merle and I started planning our wedding for early fall, between summer jobs and the start of fall classes. My family made few comments about my personal decisions, but respectful of their university-educated son, they did not question my choices, especially my choice of wife. Nor would my grandmother dream of saying anything but a warm-hearted, "Congratulations, Doug."

A small September wedding, with relatives and friends, at the United Church on the university campus near the university hospital. It was a beautiful setting for a wedding. Although, it was harvest time, everyone attended and wished us well. After a two-day break, we were back to classes and settling into a small, furnished, single-level house we had rented. Our relationship remained very much like it had prior our marriage. We were students with our own professional goals, and as friends, we had few common interests. It may have helped if we had been in a relationship for longer or if we both put in more effort, but I suspect not. Our maturity levels were much different; Merle had already experienced personal, life-altering experiences during her first years of university. I tried to relate, but at twenty-two years old, I was doing my best to put one foot in front of the other. I was only partially awake in our relationship, partially alive, and our relationship just did not mature.

As they say, sometimes in life's crazy classroom, we may choose the wrong partner, yet without question, we get the exact right kid. While Merle and I decided to part ways, I am forever grateful to Merle for giving birth to our wonderful daughter, Bettyanne, an important member of our family.

—

I purchased an old little Vaux Wagon (VW) for getting to work and classes. The car needed transmission work, which I needed to fix. It turned out all I needed was a simple a car jack and a set of small end wrenches. I had a mechanic repair the transmission, and I reinstalled everything back together. I loved cars, with their perfect containment, quiet enclosure, and sturdy wheels. They were always there to transport me to a safer spot when needed. I found myself getting more and more enthusiastic about the mechanics of my simple little German-made car. Its undercarriage was entirely closed in with a bottom plate, so the faster I hit a snowbank, the better I slid over it. Lingering left you teetering on top of the snow. Another excellent feature was the interior gas heater. I am not sure it was the safest, but in winter weather, the temperature would warm up in minutes to welcome passengers.

Model 55 HI-LO John Deere combine made by Doug

I bought myself a little home-made table saw and started doing some evening woodwork. In my Saskatoon basement, I built my younger brother, Darcy, a detailed model (to scale) of a self-propelled John Deere #95 combine for Christmas. I guess Dad and Darcy saw the value of all my work, because in 1987, upon removing the second floor on the old farmhouse, we found it stored out of site in the attic.

My love of designing and building came early in my life. As a six-year-old I would play with a small tin of shingle nails, a kid's hammer, and some wood building a toy boat or truck. As my life turned out, my business was a natural extension of my earlier interests and talents.

Year later, finally having a back yard, I found myself building an eight-foot truck camper that would slide into a half-ton truck box, a little mobile home sweet home. I had an idea, but without having a truck, I could not tell if it made any sense. I finished the outside with silver metal, a window, and a back door. Although the inside was not finished, there was a bed above the truck cab with a small counter and coolers for food. It was ready to hit the road at a moment's notice.

That next spring, I traded my car for a new half-ton Chevy truck and took a ten-day holiday through the Rocky Mountains to Victoria, British Columbia, visiting Merle's aunt Betty. A professional nurse who travelled with the provincial government, Aunt Betty lived in a ten-floor apartment, overlooking the Pacific Ocean and Victoria's harbour. I was amazed. This—my first trip west, through Alberta, to the Pacific Ocean—left a profound impression on me and provoked thoughts of possible future opportunity.

—

"Darcy, care to explore Rowans Ravine with me?" I asked.

"Last Mountain Lake? Yes, Doug!" he replied.

"I just traded my car for a new three-speed standard Chevy half-ton to carry the camper I'm building. Want to be my first VIP passenger?"

Impressed with a brilliant beach-side campground project—an astonishing water world of pleasure smack in the middle of the arid prairies—this wonderland soon became a welcome respite for hundreds of tired souls to come and dust off their exhaustion and soak up the cure: nature and fresh air. I was as excited as my kid brother, although unaware of what was brewing inside.

With my new half-built truck camper, completed in the back yard of my Saskatoon rental house, I headed to Earl Grey to pick up Darcy. We enjoyed a leisurely drive along the dirt road leading from Bulyea to Rowans Ravine, Saskatchewan's new provincial campground project, featuring newly planted three-foot tall trees, hiking trails, and sandy beaches. The pole arch with the hanging Rowans sign at the entrance of the provincial park stood out over the newly planted trees and amongst the natural prairie grass like a shrine.

With Darcy's window down as we travelled, the fresh scent of green tossed hay waiting to be baled reminded me of the hot summers. Darcy's fine hair waved in the wind over his happy freckled face. Despite our fifteen-year age difference, and the fact that our camp neighbours assumed I was Darcy's Dad, I was glad to have organized this outing. Together, we set off to the beach.

"First things first, right Doug?" Darcy said, laughing as we left the set up for later. After a good swim, I cooked the fresh eggs and noted, with some pleasure, that they remained intact in my sturdy vehicle during our drive.

Sautéing mushrooms to toss into the scramble, I watched Darcy's expert bread toasting over our small stone pit. "Enjoying yourself these days, Darcy?" I asked.

"Yep. Mom and Dad drive me back and forth to see my friends whenever I want. Donna and Bill take me on trips. We eat burgers and fries on the way home."

"Great, Darcy," I replied, tossing some baked beans into a warming pot. *What a difference a decade makes*, I thought to myself. Then Darcy got a strange, worried look on his face.

"Did you have fun on the farm when you were a kid, Doug?" he asked curiously.

"Yes, I did, although I don't know I'd use the word 'fun' to describe my life as a kid or as a teen on the farm."

"What word would you pick, Doug?"

Balancing an omelette with a serving of the grated coleslaw Mom sent along for both her sons, I considered for a moment. "I think I'd choose the word 'absorbed', yes, that seems to fit."

"What does it mean?" asked Darcy, a little boy who cared about my inner life.

"Well, a sponge absorbs liquid, right? I think I soaked in the idea that Mom and Dad were overloaded by farm work. Mom made time for us, piling a bunch of school kids into the open box truck, taking us to the annual school sports competitions for the district in Cupar and Sunday school, weather permitting, but most of the time I fended for myself."

"No wonder you love cars!" said my insightful sibling. "You made sure you got rides!"

"And," I continued, "I absorbed Dad's worry over the cost of such outings."

"Oh, he still worries," Darcy said giggling, "but Donna and Bill told me to ask them if I want anything. Did you spend a lot of time alone on the farm, Doug?"

"Well, Mom and Dad were there, and Donna, too, but yes, I spent a lot of time alone. But that is not a bad thing, you know." I hastened to assure my brother. "Every way of being has a credit and debit side."

"You mean, rewards and costs."

It washed over me, in that gentle evening light, that this conversation was genuinely brotherly—a brother-to-brother connection bridging fifteen years.

"Tell me, Doug. Did you find ways to earn extra money?" asked Darcy, choosing a fresh, juicy peach over the potato chip bowl.

"I sure did. Every day after school I washed the classroom floors and blackboards. I earned a small regular amount. At home, aside from my chores, I climbed trees to collect crow's legs from their many nests. I carried the smelly bags full to school to collect the government bounty for them. Ten cents a pair."

"Poor things," said Darcy, an animal lover like his yet unborn nieces, Bettyanne and Cathy. "But I get it. I get it," he said, nodding his young head at life's complexity."

"Yes, Darcy. Crows are by far, the most harmful bird for crops, because of the amount they eat and the area they damage. You wouldn't be enjoying that delicious peach right now if Carrion Crow had a say."

"Donna says they can grow up to twenty-one inches long. Those are big bellies to fill!" said Darcy. "Donna says they'll eat anything digestible."

"Yup. First, crows eat the grubs from the base of wet crop leaves, and then they move on to their main course: the crops themselves. I guess back in the day, when moderate-sized family gardens were protected with a scarecrow or mesh, things were different. But today, the large commercial vegetable and fruit crops grown to serve a nation of people need different protection. I learned in Boy Scouts how to climb up a tree and tie great knots. I especially loved the winter months as a farm kid when I had more free time. So yes, Darcy, 'absorbed' would be the word I'd pick, because I was, without knowing it, wholly engaged with nature during that free time on the farm."

"Wow," said Darcy.

"I remember when a classmate couldn't believe I didn't know about Elvis Presley!"

"Who?" asked Darcy.

"A famous singer," I replied, smiling. "Like the Four Tops or the Beach Boys today."

"What else did you do by yourself?" asked my younger brother.

"Well, I had a trap line for snaring and selling rabbits. The snow path averaged three miles when not blown over by storm snow. Occasionally, wearing my trusty moccasins or the brilliant Native invention of snowshoes, I'd trap the odd weasel."

"All alone?" asked Darcy.

"Yes. I did my game hunting on foot. Sometimes, though, I'd ride my trusty black pony, or I'd harness him, and he'd pull my cutter. Bush rabbits were plentiful at the time, and I'd trap twenty on a single trip. I sold them to townsfolk for pet food. Gopher tails were also worthwhile, but I never mastered the technique of snaring them."

"Donna says they're speedy critters," said Darcy. "Aunt Jean said you sold her the garden seeds you ordered from a magazine, and Mom told me you sold wood from our pile for a couple of years. Wow...I think you landed on the reward side, right, Doug?" At eight years old, my little brother was astute.

"I kept my eyes peeled," I replied, tucking Darcy into a guaranteed warm sleeping bag—Canadian made. I don't know which of us fell asleep first.

"That's a sharp-tailed grouse calling, Doug." I caught his voice at 5:30 a.m.—the two sons of Kenneth and Hazel, flat on our backs in our tiny stowaway camper, never mind our sprawling three-bedroom farmhouse only miles away.

"Also known as a prairie chicken," I added.

"One of Saskatchewan's most popular game birds, Doug. Do you know what year it became our province's official bird?"

"No, I don't know, my learned lad." I smiled again.

"1945," I heard, and then nothing from either of us for three more hours of solid sleep.

"Donna said Dad built dog crates for you, Doug," said Darcy over our pancake breakfast. "You'd both take care of them until you found a buyer. Then you and Dad would load the dog in his crate onto the train and send it off."

"Yes, and once a puppy arrived at its new home, the crate was returned, and I'd pick it up at the rail station. I would save the money I earned for special occasions, like the sports day in town or the Canadian National Exhibition in Toronto when we travelled to Ontario with Mom. Earning my own extra money was my failsafe."

"What did you do after school, Doug?"

"Well, before our first TV, I'd spend time with your doting sister Donna, or by myself, skating on our shallow frozen pond. I can still hear the clear ice cracking under my skates. Sometimes I'd jump my horse over obstacles or build rafts for summer swims in the dugouts. That was fun, too," I said. "It was pure enchantment for me. I got absorbed into another world."

I noticed Darcy's head hung a little. "I wish we could a' played together, Doug. I wish we could have' grown up together."

"Me too, Darcy. I wish that, too," I said, suddenly wondering where I put that bag of marshmallows. "But our birthdates had nothing to do with either one of us, right, Darcy? We're born when we're born," I said, noticing Darcy's lowered head. "Darcy, look, let us make the best of it, okay? Let us consider this visit as part of our growing up together. We are having a great time today. And I wouldn't have been able to tell you about my farm history if I had not lived it already, now would I? And who knows, with a big brother like me, I can look out for cool opportunities for you, okay? Maybe I can get you jobs in the future, same as Dad did for his younger brother back in the day. And remember, Darcy," I said, pointing to my vehicle, "I have wheels. Big enough for the two of us!" Ah, that got a smile, one I loved.

"Anyway, Darcy," I said, as we prepared to explore Rowan's Ravine, "We can find plenty of fine brothers and sisters in this big old world, right? I challenge you, Darcy! Choose good friends and head out to the fields to explore. And next summer, we'll sit in this same campground, and you can tell me about your adventures, okay?"

"Deal," agreed Darcy, placing smaller rocks around the fire pit in preparation for our after-swim cookout.

"Do you know the meaning of the First Nations word "Saskatchewan," Doug?" he asked.

"No, my teacher. Educate me."

"It's about the Saskatchewan River. Donna said the Cree named it 'the swiftly flowing river.'"

"Let us go then," I said, "my brother."

—

A full month later, in 1965, alone in the farmyard when my first-year law grades arrived, I slid a blade along the emblem end of the University of Saskatoon envelope. My fingers begin to shake. I did not get a passing grade in two of first-year law courses. I held the report—my hands sweating, my heart pounding. What happened? I thought I had done better. Shocked, I grasped for options. Did it have to do with Merle? She was touring Europe with Almina. Crushed, I sunk into the back-porch steps. Still, within minutes, I'd arranged a meeting with Dean Otto Lang for the following morning at 8.30 a.m. sharp.

"You're young, Doug! So young," said the Dean of Law at the University of Saskatchewan. "Time is on your side! Simply redo the first year. That way you'll be well-equipped to complete your law degree." Dean Otto Lang was the youngest person ever to be appointed to his post at the University of Saskatchewan, and he had a distinguished career in politics and business.

As I absorbed his quick solution, the blue bruise I'd been feeling since my wedding deepened into a mottled purple. Later that week, I received a postcard signed by Merle and her friend Almina. Together, they had visited the British Isles, France, and they were now in sunny Spain. I began to work on finishing the interior of my truck camper. First, I removed the double handles and two latch parts for the trailer door from their twin package. They did not fit—misaligned. They reminded me of Dean Lang's speech about personal autonomy being hard-earned. I reviewed the facts once again. What about a career as a commercial lawyer? Could I pick up the extra classes and redo first-year law?

I parked the car outside my barbershop. It was time for a trim, but while I sat there, the voice in my head kept talking.

You enjoyed your fourth year with all the economic and senior accounting classes. You did great in the final exams.

The barber snipped a new hair style, one I had not worn before. I sensed change.

I have been in post-secondary education for five years, I thought. Looking in the mirror, I assessed that I would be staring at five more years of school, if I attended law school for three years followed by two years of articling. *Forget about income*, were the last three words I heard in my mind as I watched the barber masterfully wield his scissors. At twenty-three years old, I had a huge decision to make.

—

It is funny how hunting and fishing with friends helps settle one's mind for decision making. In first-year University, Jerry and I hunted Canadian geese north of Saskatoon one weekend. The first time I had done this was with Grandpa's old double-barrel ten gauge shotgun that Grandma gave me, which I later lost. The owner of our rental house let us use his garage to pluck the birds and cooked up a big feast for everyone in the house. Dad and I were not hunters, although I did a lot of hunting on our farm during my school years. We mainly hunted small game. Once, a couple of neighbours and I roasted a pigeon we'd caught, and Dad had cooked a rabbit for supper that tasted very good, like chicken.

A few years later, when Jerry was teaching in northern Saskatchewan, we got to experience the nature there, with the creeks running into the bountiful lakes, which offered northern pike fishing. One early morning, a little northeast of Tisdale, we dropped his boat off in the shallow sand and rock beds close to the wide-open lake some ten miles away. He advised that we walk alongside the boat and nudge it along with our legs until the water deepened. Water levels and current speed vary depending on the time of the year or during rainy periods. This day, the current played tricks and led us away from our destination. Suddenly, the creek widened into an idyllic scene that took my breath away. Before us spread a tree-lined shore dotted with two old log cabins and beaver skins drying on racks, their lines strung high so that no wild animal could relieve the trappers of their bounty.

Arriving at the mouth of this unnamed lake, the sky blue and breeze soft, my shoulders relaxed, and I began to breathe deeper. Our boat rocked gently on the waves as Jerry and I reeled in one weighty catch after the other. We fished all afternoon, and silver swimmers weaved in and out along the weed beds. What fun we had.

Later, after fileting our fish and setting up our sleeping bags in the empty trapper cabin, we lit a fire and enjoyed one of Jerry's great fresh fish suppers. We spent another great day at this beautiful spot before heading home. An intense, much faster current surprised us on the journey home. Our shoulders up and breathing clipped, we plowed through the shallow waters, lifting the boat motor so as not to damage the propeller. Then, back in the deeper, fast-moving stream, we lowered the motor and started it using the pull rope. The paddling, getting in and out of the boat, and working the motor in the current needed two things: a calm demeanor and experience, which Jerry quietly demonstrated. Although lacking experience and patience at times, I understood that it was all hands-on deck and I would follow a knowledgeable leader, working hard for the cause. Hours later, we were tired, and it was almost dark, but our transportation home was in site. I expect we toted home more fat fish fillets than today's law would allow.

Heading home from nature's classroom that day, I began to inquire more than I'd ever permitted myself to if I was enjoying my life. On that drive, I began to reconsider both my personal goals and my career goals.

Out of nowhere on that drive home with Jerry, like a surprise bobcat on a stormy lake, Mr. Pinkney, a homesteader I'd never met, appeared beside me in the passenger's seat in a tattered, poorly cut coat. Long ago he had built the older house and the historic beauty of a barn my parents bought when they returned to Saskatchewan in 1949.

"Look, Doug, there's no straight line to certainty," said

Mr. Pinkney as though he had known me all his life. "Your own parents went back and forth, right? They auctioned the farm, sought a better life in Peterborough, and within a year back they came.

"Yes, to their new farm and the sturdy house I built in 1909," Mr. Pinkney continued. "That barn was not the one I'd ordered, Doug. They shipped the wrong design. Material for an eastern style square flat-roof arrived in rail box cars, not the western hip roof style I had requested. But the wood was quality, and with winter coming, we built it anyway. Some years earlier I'd also up and moved off my Ontario farm. We herded our livestock and hauled our belongings on foot. Step by dusty step we trod to the northern US where, very soon, we made another choice. We moved to Earl Grey, Saskatchewan—a homesteader's dream. For this third move, my shoe leather worn thin and we travelled by rail. I lost all I had built so far, but not really, Doug, because at last I owned something far, far better. I now possessed a quality called confidence.

"Mrs. Pinkney was proud as punch of her new Earl Grey house with the modern pantry bin that held a full one hundred-pound sack of flour. She was thrilled with her modern porcelain wash sink and green metal hand pump right there in the kitchen, with an attached cement cistern under the kitchen floor. So I too had to go backward to go forward, Doug. I too, got the wrong design at first. But over time, I figured it out."

"Yes!" I reply, startling Jerry who is enjoying a wee nap on the passenger side during our drive home from our great weekend of fishing and fun. In retrospect, my career choice might have been more suited to me. A confined office job in a law firm would not have fit my life, just as being one of the hundreds of factory workers in Ontario was not the ticket for my Dad. Yes, I enjoyed preparing for the mock trials in the law program, especially improvising on my feet or belting out salty retorts, sharp and on the money. I might have done reasonably well, but we will never know. The good news, and this became my litmus test for all my decisions, was that I felt no regret. I never looked back.

That evening, after a stop off at the hardware store close to home, I attached a new single brass handle on the trailer door and threw it wide open with a force that surprised me.

BECOMING A PROFESSIONAL ACCOUNTANT

"CONGRATULATIONS, DOUG," MOM SAID HAPPILY. "AT AGE twenty-three you are starting your professional career with an outstanding national accounting and audit firm: Touche, Ross, Bailey, and Smart."

"Thanks, Mom. I have made my decision. I have signed up for a three-year chartered accountant program. I'll do my articling in Saskatoon in their downtown office."

"And study at night? Wonderful. What will that give you?"

"My Chartered Accountant Certificate (CA), Mom, and a salary."

"You remind me of your Dad and Uncle Calvin, Doug. They worked in Peterborough during the war and studied at night, too. You have got your life organized, Doug. Good for you."

—

That summer I began my new job as a student of professional accounting. There were no computers or software programs in use yet, thus information and audit files were handwritten or typed by a secretary. It is hard to imagine, now, how a necessary change or error on a client's file required hours of adjustment, compared to today's push of a button. Even after all the accounting, economic, and administration classes, I had a lot to learn in the audit (verification and financial) reporting business. Audit is verifying—checking financial control and asset value for clients and owners. Only the chartered accountant's signature (now known as the chartered public accountant) certifies and approves to be true and correct any audited financial statement. As an articling student, I gathered information, verified values and quantities, and assessed accounting controls and management procedures. I was good and loved doing it!

In Saskatoon, I worked on municipal tax-payer reporting, recording financials for Federated Co-op's annual statements, helping set up a public company selling shares on the stock exchange, and preparing a variety of financial assignments for many other companies in and out of the office. The firm's obligation was to prepare their students for the national chartered accounting finals, which meant three years of articling in Saskatchewan, at minimum wage. That first fall study course and company policy sessions were provided in Toronto by the firm's national offices. Two expense-paid weeks in a big eastern city meeting many high-caliber executives sure impressive this farm boy.

Toward the spring of 1967, I began to consider transferring to the Calgary office of Touche, Ross, Bailey and Smart. I had been through Calgary on a holiday when I went to Vancouver Island. That was the first time I had seen the mountains, and I was impressed with the city's proximity to them. Also, Alberta offered a two-year articling period if students already had a university degree; that would save me a whole year of additional articling time, if I could pass the exams. In June, the firm sent me for an interview, and by fall, Calgary was to be mine and Merle's new home—Merle finding work at the Calgary General Hospital pharmacy.

I traded my Chevy half-ton and camper for a new green two-door 1967 Ford, to be picked up in Toronto before travelling on to the 1967 World Expo in Montreal. Aside from the French-speaking attitude, which I was not prepared for, Expo 67 was great. The move to downtown Calgary followed. We rented an apartment in a downtown high rise close to my office, becoming Alberta residences in the growing cowboy city of Calgary, only fifty miles from the Rocky Mountains.

I settled into my transfer and immediately began studying. It was a welcoming office, as Saskatoon's had been. I learned a lot. I developed new friendships, particularly with Ernie and Harvey. Ernie grew up near Rosetown, Saskatchewan and lived close by, so we spent a lot of time studying together. Harvey already had his CA designation, and his wife, Sharon, hailed from Manitoba. Over the next few years, the three of us enjoyed two-day weekend trips to the mountains to fish and camp. The final University of Calgary exams were soon upon us. Critically important, success meant a substantial salary increase and new opportunities. The success record was usually only fifty or sixty percent, so there were no guarantees. Six exams over six days was stressful for me. The examinations divided into six main areas: the Canadian Income Tax Act, auditing and its rules, professional standards, accounting, and business administration and asset valuations. I was ready for the hardest set of exams I'd ever written. I had good summary notes and even a bit of a flair for this commerce thing. Dad always said we needed a bit of luck, too.

When the marks were posted, I learned that I had passed, and so did my buddy Ernie. What a great feeling! I was so glad to have completed my studies. After seven years of schooling, I was ready for the real world. Harvey and Sharon threw a party for the office that night, and we celebrated and celebrated. Immediately, the world changed for me. I was ready to make the most of it.

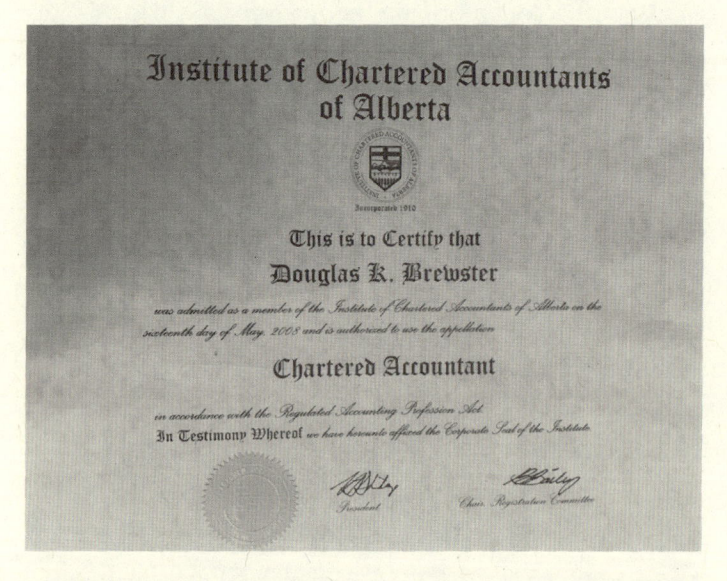

In a large firm like Touche, Ross, Bailey and Smart, there seemed little difference between the work of a senior student and a new CA, except in the degree of supervision on the job sites, more responsibility for completing files for the clients, and of course, a salary increase. Many graduates moved into the business world, for additional money and opportunities, while others stayed with the firm, aiming to become a partner someday.

A Lucky Prairie Boy

In 1969, the firm became known as Touche Ross—far from its humble beginnings in 1845 by Mr. William Welch Deloitte in London, United Kingdom. Today, it is called Deloitte, a multinational firm and one of the largest professional service networks, involved in all aspects of the business world. I am proud to have trained under their guidance and leadership, and I continue to remember those important years of my life.

During this time period in Calgary, I did the financial books part time for small business clients. One client, Mrs. Lee, owned a travel agent business as well as property in China town, which comprised almost a full block on Centre Street next to the Bow River. Her husband had recently passed away, and with her only son in Toronto, Ontario, Mrs. Lee needed financial help. She also referred me to a couple more local Chinese businesses. I also did a lot of work for a contracting company, Unruh Construction, owned by Wilma and Peter Unruh. Their construction company built most of the Calgary A&W outdoor restaurants, where waitresses took orders and delivered them on trays that hung on car windows. I spent many hours over several years doing their accounting records.

I transferred to the bankruptcy department with Touche, Ross, Bailey and Smart for a short time before leaving public practice and the accounting firm. This work gave me a different prospective on owning a business and the risks involved in the free enterprise system, which was helpful in my future years. Showing up to a shoe store, having the locks changed to keep owners out, counting inventory, and seizing assets for the creditors gave me a better understanding of the financial control in the business world. It was all about the money.

The first house "owned" in southwest Calgary, I bought with very little money. The previous owners were willing to sell it to us under an agreement for sale and carry a loan with us, thus retaining ownership of the home until it was completely paid for. We welcomed our beautiful daughter Bettyanne Louise into the world, and into our first home in June 1971. Merle continued to work, so we hired a wonderful lady name Audrey to look after Bettyanne full time. She did an excellent job, especially in those first few critical years.

The second house I purchased was in northeast Calgary. In this house, I built and furnished a basement rental suite. The objective was to make extra income, so the house would eventually pay for itself. Building that suite was my first big project. I worked through Christmas season completing the framing, dry walling, and mudding. My friend Ernie helped until he slit his arm with a utility knife and I had to rush him to emergency. I was embarrassed about the job on the drywall joints; I put way too much product on and could not sand it all off. Every joint was visible, especially in the ceiling. I furnished the suite with cheap new furniture and successfully rented it to a couple from Saskatchewan.

—

"Harvey? Ernie? Is that you on the line?"

"C'mon, Doug, let's go!" said Harvey. "Our pilot's waiting! I got us a small four-seater." I jumped into the small plane and slammed the door. Goodbye to work, for two days. I fastened my seat belt and turned to see our fishing rods, sleeping bags, gear, and cases of beer, food, and water. Minutes later, we taxied back to the hangar. We had to lighten out load, so out went some of our gear, but not the beer. Then lighter, we were flying north at a reckless speed.

"I'm a bit of a stunt man," the slick young pilot said with a wink at around six thousand feet. I spotted the airfield below. Spiraling down like a bottle corkscrew onto the runway, we gasped at his show-off landing. Reeling on the grass landing strip, our pilot relished at his terrible joke. A truck then rolled up to drive us to our cabin, which was a little log chalet that had seen better days, located on a riverbank mountain Native Reserve.

"You filet the fish, Doug," said Ernie. "You're the only one who knows how to, anyway. I'll tend the fire."

That weekend of fun, fishing, and a few drinks with pals was just what I needed after the long days working as a professional accountant and nights working as a carpenter

A third house was purchased in southwest Calgary. I rented it to a young woman, Ann, her two German shepherds, as well as Almina, Merle's friend from university. I soon became friends with Almina's family and rode horses and hunted geese with Almina, her sisters and brother at the farm, south of Calgary. During those years, I misplaced Grandpa Brewster's old ten-gauge American shotgun— the one I had used to hunt geese in Saskatoon during my first year of university. I think I left it at the Stavely farm while hunting ducks, with young Bill. He was too young to hunt on his own and I must have forgotten to pick-up later.

Almina and her two sisters, brother, and parents were great to me. The year I worked on my first basement suite, they invited me for Christmas dinner at the farm, which I gratefully accepted as Merle had gone home to her parents over Christmas. After Merle's and my separation, I did not want to cause any conflict for Bettyanne and her mother, so did not continue visiting them.

I had started to build a suite in our third house, so I dug a big hole in the cement floor to install the new bathroom pipe, but since the girls rented the whole house, I wasn't able to finish the job. A few months later, a water main break on the street flooded back the large hole and swamped the basement with three feet of water. Silt damaged the lower level, but luckily for me, my insurance took care of every repair.

In November 1969, I took a job with a small manufacturing company called Flagliner Industries in downtown Calgary. Fred, the owner, had been advised to hire a chartered accountant to help straighten out his financial and accounting issues. This was my first exposure to a manufacturing plant's planning, material processing, assembly, painting, and cost control. The offices were small and dirty, but I quickly improved them to create a better work environment.

The company had a variety of products for the building industry and manufacturing of trailers. Fred had designed and built a gooseneck type freight trailer where the hitch was centred over the rear axle of a half- or three-quarter-ton tow truck—I was impressed. His impressive new concept offered a more economical way of moving smaller and lighter freight, as opposed to having to use larger semi-trailers. I got my first experience pulling one when I delivered a twenty-four-foot unit, complete with cattle racks, to High Level, Alberta.

A couple homesteading north of Grand Prairie purchased one of Fred's gooseneck freight trailers for their new farming venture. One weekend, I took it up to their farm; however, I never saw their actual homestead due to muddy impassible spring roads. The young energetic couple, both with university educations, met me a couple of miles from their farm. It was a step back in time for me; the bush, snow, and mud reminded me of my early days in Earl Grey, particularly when we had to make the necessary trip to Earl Grey early in the morning while the ground was frozen solid.

Fred's manufacturing business was suffering with past warranty costs and needed more capital, which was common in small, private business ventures. I began to worry that my salary was an additional burden to his financial situation. Reporting bad results and making future projections for additional profits without further funding was not working. In retrospect, with more experience I may have been able to help him more with his situation.

That May, I noticed a job offer for a financial controller with an American company in their new, soon-to-be-built mobile home manufacturing plant. It was to be in southeast Calgary and have an operation of 250 employees. I applied, receiving a call from the general manager, Bill, who invited me to supper at the Husky Tower, now called the Calgary Tower. Bill was a friendly fellow with an engineering background in the automotive manufacturing industry. He had just been hired and transferred to Calgary by Commodore's head office in Omaha, Nebraska, USA to manage and set up a new 110,000 square foot plant. They'd already started building mobile homes east of Airdrie, Alberta, in an old-World War II air force hanger. As luck would have it, I was offered the job and accepted without delay. Jerry, my friend from Saskatchewan, was not happy when I canceled our planned trip to Alaska that summer. I felt terrible to have let him down, but the job would not wait for our return. We never again got the chance to make that amazing trip up through the Klondike on that beautiful scenic highway built out of necessity during the World War II effort.

I began working for Commodore Mobile Homes in June 1969 at the Airdrie facility. There were six office staff and about fifty plant employees at the time. There had been an accountant in charge of the setup, but with a young guy like me overseeing him, he soon left. I was on my own. The temporary office being used was small and not soundproof, with only four desks. We operated out of this space while the new plant was under construction. Bill, the general manager, handled the coordination with the general contractor and equipping the new plant, as well as running a start-up operation. I played a role in the 4500 square foot office layout, which was in a single-storey building attached to the central plant. The plant was completed that fall, and we moved both the office and plant staff.

The administrative functions and staff were divided between operations, such as production, sales, and purchasing. Finance, accounting, and cost control were my responsibility. I reported directly to head office in Omaha, Nebraska. My work here was quite an education.

—

I really enjoyed the outside distractions of fishing and camping that came with living in Calgary. But in the early 1970s, there was little time for such recreation. I had purchased a twenty-acre acreage just east of Airdrie on Big Springs Road, near Commodore's original temporary location. My friends, and even Dad and young Darcy one weekend, helped me build a little cabin featuring a beautiful view of the mountains. I never had the time to do more than a bit of landscaping and small renovations. Friends Ernie and Harvey visited, and we'd celebrate until the wee hours of the morning. My acreage became a central gathering point for friends from Commodore. I purchased a new twelve by sixty foot mobile home for $6100.00. I bought it right off our production line, built a garage, and stacked scrap lumber from the plant to use in our fire pit. A sparsely populated area, my place was a great gathering and relaxing place, which I enjoyed. It is now located within Airdrie's city limits.

—

Commodore's manufacturing plant was designed for purchasing raw materials and processing them within the plant. Exceptions were furniture, windows, doors, and appliances, all of which were transported by rail from suppliers. We had an unloading dock large enough for three rail cars inside our plant. Cabinet making and painting, as well as making the metal roofs, sidings, and steel frames had a separate area equipped with specific equipment for processing and building. The main production line had eight work stations that had to move forward simultaneously to be efficient.

The workflow (and stations) were as follows:

1. Build steel undercarriage and install axles and tires.
2. Build wood floors, linoleum, and carpet and mounted them on frames.
3. Build walls (interior and exterior) on a jig (also known as a table) and install on the floor. Install glued and stapled paneling on one side only.
4. Rough in of electrical and plumbing.
5. Install remaining paneling, molding, and cabinets

6. Install sheet metal siding, trim, and roof.
7. Install plumbing fixtures, electrical fixtures, doors, and windows.
8. Install furniture and appliances, clean up and quality control inspection.

The production line was timed. Everyone moved in sequence. Stations one to three moved straight ahead, four to six moved sideways on rollers, and seven to eight moved forward to the exit door. Many of our supervisory staff had experience in the automotive industry. They knew that the number of units produced in a day depended on how many workers were assigned to each station. Production function and incentive performance were reviewed continuously. Daily efficiency reports were prepared, with weekly summaries forwarded to the head office for their review. The highest production in a day was eight new units out the door with around 250 employees on the production floor. Our engineers modeled our production system based on automotive production systems.

Below is an example of an efficiency report.

Floor decking:
1. 24.18 minutes budgeted (projected)
   27.6 minutes actual work
   88% efficiency
2. Metal skin on the exterior:
   9.71 minutes budgeted
   8.0 minutes actual work
   121% efficiency

Wages per hour were negotiated with employees, plus they received efficiency bonuses when their time worked was less than time projected to complete the work.

A Lucky Prairie Boy

Huge efforts were made to capitalize on the new and expanding Canadian mobile home industry and market. Bill and his team, me included, were encouraged in our aggressive effort to be part of this growing market.

I reported on daily production, manpower, unit delivery, and new sales to head office, along with supervising the accounting department and its various functions. Trade shows, dealer meetings, and company functions were part of our everyday activity. We had a great comradeship and became a close-knit group.

Commodore Management Team

Accounts receivable and dealing with delinquent accounts across the Prairies was challenging. It was not all about simple reminder phone calls. One time, Al, the vice president of sales, and I flew in a small chartered plane into Hinton, Alberta—a big logging and mining area. With more experience he wanted to show me the way to settle unpaid accounts and problem dealers. A strong wind on the icy runway blew our plane close to the side of a snowy mountain. Our eyes widened as we saw that a prior plane landing had careened off the runway and plunged into the snow and bush. I was dressed for the winter, but Al was wearing only a fall trench coat. He shivered badly while we waited for our taxi at a lone phone booth with smashed out windows. When our business was done, we flew back out in bad weather and the wing started icing up. Our pilot had to change course and fly low to prevent icing. He flew along the highway in case of an emergency landing. Al, the California native, declined any further accompaniment with me into the north.

During my second year with the company, Commodore offered its Canadian dealers a reward trip to Spain. An Air Canada charter flight, featuring superb meals, liquor, and partying for the eight-hour trip from Toronto to Malaga, Spain took me a full day to recover. A representative from the travel agency organized the trip and helped ensure everyone had a good time. Spain was my only venture to Europe, and it was fantastic. The mountains, caves, castles, and day trip to Morocco, along with the private dining for our group of approximately 150 people, was the opportunity of a lifetime. Such a trip at today's costs would be unaffordable for most. In 1970, a chartered round-trip flight cost $20,000.00. I will never forget the fantastic sun rising in the east over the dark shadows of the European coastline as we approached our destination, leaving the Atlantic Ocean.

With no computers or memory sticks yet at that time, Commodore's system comprised of a one-write system. This was a basic manual system with a special board and sheets of carbon paper to cut down duplication and error, saving time.

By default, again, I learned more about my natural and unnatural talents through my work here. One day, I did the payroll for the absent payroll clerk. Soon enough, half the staff rumbled into my office complaining about missed hours or mistaken deductions.

Meticulous detail and daily routine was not my strong suite. It was also why the phrase "he needs his people" took root in my life. The next day, though, I negotiated a global labour contract with the union, in favour of our employees, and emerged looking great. I guess I'm a guy who needs a challenge.

After the operation was established, issues with the quality of our product began. Units had been sold and were being lived in across the country, including on Vancouver Island and Terrace, British Columbia, and east to Brandon, Manitoba. Bill was responsible for both quality control and quality control personnel, including hiring the necessary repair service people across western Canada. The head office, however, was not prepared to wait for improvements and quickly shuffled us around. Bill was transferred back to the US. The vice president hired a man named Norm, from Lethbridge, Alberta, to run the Canadian division, which comprised of three manufacturing facilities: two located in Alberta and one in Ontario. I was promoted to general manager. With tremendous support from the office and plant staff, I moved into the corner office and hired a new controller to replace me. I was enthusiastic and felt confident that I could handle the difficult situation. I soon learned that job security for a general manager in this industry was nonexistent; it is all about the numbers.

The general managers' meeting in Los Angeles was an eye-opener for this farm boy. Scalpers sold us tickets to "Hair," which was playing in Hollywood. Afterward, I declined an offer for a starlit midnight yacht cruise and chose to get some sleep. The next day, my early morning presentation was well received. I spoke about sales agents wanting too much financial support rather than earning the sales incentives offered.

Dozens of trips to the United States and Ontario that year meant travelling nights and weekends. Beyond exhausted, I learned how wearing business travel can be.

The construction of factory-built mobile homes in the early 1970s was much different than today's unit construction. The 1970s mobile home was more of a glorified travel trailer, not built for Canada's sever weather conditions. Today, Canada's strict, top-quality building regulations require a much higher standard of quality than in the 1970s.

While engineered to be structurally sound, the 1970s mobile homes were built with the minimum quality material. Because of the use of glue and chipboard floors, as well as no all-weather windows and doors and little insulation, moisture and cold became the enemy.

We had two service technicians: Morris, a good family friend, and Al worked on the road doing warranty work. One day they approached me about building and selling an adjustable jack, in place of wood blocking, to support mobile homes. Al had them built and made a brochure, and hundreds were sold over the next few years.

Commodore's head office was high on responsibility and low on creativity. While our sales were good, we continued to suffer from poor quality, as well as not having the right product for the Canadian market.

Oversight on production design and construction procedures was done in Los Angeles by the American personnel at the West Coast Division. This created problems, as the Canadian market was different from the American market in many ways, including weather and culture. I believe this had a lot to do with Commodore's lack of long-term success.

My objective was to improve the quality and expand the dealership network in western Canada. As a controller who worked with the business on financial matters, I had a good relationship with and knowledge of our dealer organization. Our sales staff department was excellent, and with my over-sight and trips to the dealers, our sales were strong. I especially liked travelling to Vancouver Island in springtime. A ferry ride from Vancouver to the island, where the formal dining rooms still existed, was delightful. So was touring the island, which was new to me. There were also many new mobile home park developments in places like Ladysmith, Princeton, and Yellowknife.

On another trip to Princeton, British Columbia, where a large copper mine was being developed, we rented a single-engine Cessna out of Calgary to Kamloops, piloted by one of my salesmen. From there, a developer with his own twin-engine plane flew us to Princeton and back, hoping to make a deal to purchase one hundred mobile homes for his park. I had never flown across the mountains in a small plane and soon realized the risk of flying at the same altitude as the mountain peaks. Heading to British Columbia we flew above the valleys over Highway #1. Close to Canmore, our engine quit. I stayed calm while the pilot frantically pushed buttons to restart the engine. We lost altitude quickly. Finally, the engine came back to life. My pilot quietly commented, "Guess I forgot to turn on the other fuel lever."

On the return flight, my trusty pilot decided we could save time by flying directly between the mountain peaks. It should have been a beautiful scenic trip, but instead I focused on finding landing spots, just in case. We made it home in one piece.

Product quality was hard to deal with, as I had little control on product design, material input, or production management.

Norm hiring one of his friends as the production manager did help, and improvements could not be made fast enough to overcome our start-up quality difficulties.

I received positive reviews for our production and sales results from the top management at the head office and was always treated with respect at the manager's meetings for my knowledge and expertise.

As mentioned earlier, well before I joined Commodore, I'd been involved in Fred's small company, which was one of the first businesses to design and build a gooseneck trailer. The design fascinated me because of how much more weight a half- or three-quarter-ton truck could pull. In my spare time, I began to design and build a prototype of a twenty-foot gooseneck freight trailer out at the acreage, where I had a small workshop.

Using my employee discount, I purchased a few of the items I needed from Commodore. I never told any of my managers about the prototype I was making, as it was just a hobby.

I became friends with Rick and Margo through my work at Commodore. Rick was interested in helping me with my trailer project and purchased a new red half-ton Chevy truck to carry our building materials and pull the trailer. He also found a couple of skilled friends to help us build the prototype. Soon, a ball was put in the centre of Rick's truck, and the first unit was ready. We hooked it up one weekend to take it for a test drive but were disappointed because, after a few miles, the gooseneck arms spread apart. A simple brace was welded between them to solve the problem. We were ecstatic and took pictures of our blue, newly named Brewster Trailer Manufacturing. There was no thought of it being an affront to Commodore, so enthused, we took photos of it outside our Commodore offices. Excited with our prototype, we considered a second one. At this point, I had not thought of leaving Commodore; all was well. Fate, however, landed a big surprise.

My boss, Norm, arrived one Monday morning and declared a change of general managers. He announced that my services were no longer required but provided no explanation. Upset, overwhelmed and confused, I left immediately. I soon learned that Rick and Lloyd were also let go. I had hired Lloyd as a sales coordinator and knew him to be an excellent employee. Many years after this sudden blow, I learned that Lloyd had sued Commodore for wrongful dismissal and won. *Why didn't I do that?* I asked myself. Because I had no inkling I could. Nor did I figure that these hardworking guys were singled out only because they were friends of mine. I spent hours agonizing about that shocking dismissal—a dark moment in my past. Was it my fault? Did Norm suspect I had taken materials without paying for them? I was so proud of my work for Commodore, where until that moment, I had only ever received praise for my work. Norm, my direct boss, had not hired me. While we worked well together, we had distinctly different personalities. Soon after my abrupt dismissal, a friend of Norm's was given my position.

In retrospect, I sense that someone got wind of my design project, an innocent personal venture, and reconfigured it into an act of heresy or betrayal, as though I would high jacked Commodore's top-secret ideas. I had not hidden my project and had no intention of leaving the company. Why wasn't I questioned about my project? Found guilty without a trial, that night I found myself travelling alone on a poorly lit bus back to Saskatchewan, a trip I'd not taken since university. By the time I returned to Calgary a couple days later, I had decided. I'd represent myself. Hello, Brewster Trailer Manufacturing!

As fate would have it, I never worked for anyone else again. My dismissal turned out to be a lucky break for me, as was being hired in the first place. I was sad, though, that my friends got roughed up. But Lloyd, a loyal employee, found work in related sales, and Rick came full time in the brand new and unexpected venture of Brewster Trailer Manufacturing. Life threw me a nasty curveball.

PART TWO

BUSINESS VENTURES

I HAD JUST STARTED MY OWN BUSINESS, but now needed a salary. With only two employees doing the fitting, welding and assembling the trailers, I took a production shift completing the trailers, with Rick doing sales and securing materials, along with helping me finish the units.

Our little shop at the acreage east of Airdrie was just big enough for one unit and had little room for storage. Without a proper driveway, the rain and spring thaw blocked our access, so just as had occurred on the farm when the unpaved roads were thick with mud, we waited until the ground froze to move units and materials.

We heated the shop by a makeshift woodstove. My job was to deck the unit and paint it during the night shift so that, in the morning, it could be moved outside, and another frame could be welded. We had no idea about the future. We only had a plan, our optimism, a quality product, and a market demand. Once again, we got lucky.

One day, after several years of spending too much time working, travelling and visiting all on my own, I asked Merle to be clear with me. "Tell me to pack my things and be on my way. I'll respect your request." At her brief nod, I packed my travel bag and left that night. I stayed in a Calgary hotel for a while, and then later with friends, finding longer term arrangements. That door finally closed; my marriage was over. Visiting arrangements for Bettyanne took some time and involved lawyers. The financial arrangements were easy, because of our existing independence. The joint house was transferred to Merle, and at last, my light breathing was restored to deeper inhalation. With the encouragement of my friends, I swapped the judgemental words "failed marriage" for the more compassionate ones: "learning experience." From there, dozens of doors began opening for me, one after the other. Life, forever renewing, was supporting me.

—

In the beginning, I drove a green two-door Chrysler that I'd received from Jerry in 1972. He had purchased it new and then decided he'd take my 1967 Ford on trade, which I'd bought new in Ontario while visiting Expo 67. We met to switch cars in Earl Grey at his parents' home, where I enjoyed a visit with Jerry and his sister, Janie, an interesting young woman whom I had known from high school. Although I never preferred Dodge or Chrysler products (I'm a General Motors guy), I purchased a number of them over the years. The Chrysler transported materials when Rick and his truck were not around. It was our first official company vehicle until it was traded for a shiny new orange half-ton Dodge.

Soon I rented our first manufacturing shop in Calgary. It was a shabby old building in the northeast area. Now, we could double our production with no night shift. We built a portable office and placed it in front of the work bay, added an entry door into the shop wall, and voila: our shop office!

One day, while working in the yard, I noticed my former boss, Norm, and my former US boss, Leonard, scouting my bay area. They did not stop to greet me. I let on as though I did not see them, but I often wished that Leonard—a man whom I admired and had thought a lot of me—had called me to discuss my abrupt dismissal. I am sure the story he heard was far from the truth.

We never crossed paths again. Still, those intense years in my late twenties equipped me with essential personal tools. If I could go back in time, I would have picked up the phone and called Leonard the day I was wrongfully dismissed. We could have talked. I'd have used my voice. Of course, he could have called me, too.

We soon outgrew the space and rented a second facility close to downtown Calgary, on 9th Avenue. The new offices were far larger than our little twelve by twenty-four foot portable unit in Forest Lawn. We shared the building with a new appliance storage business, and just behind the office building was another long shop that was perfect for manufacturing wood products. The welded frames with axles and hitches were built in our welding shop and then transferred to this location. Eventually, I hired a second secretary and a draftsperson.

My idea of producing smaller portable buildings began a few months later. The five guys employed in the shop produced two twenty-foot trailers a week. When Rob joined us, his attempt to build a small six by eight foot mobile home porch on the grass out front of our office told us he was not a carpenter. He installed the asphalt shingles, intended to prevent moisture from entering, top down rather than bottom up, thus the edges caught the water rather run off. Still, he was a good worker and supervisor, and he easily fit into our business in production, sales, and the collection of accounts receivable. That is, until he turned us upside down.

We started the second line for small portable buildings with a frame, hitch, and wheels under the unit. Quality product was built and shipped. Jerry came to work with us for the summer. As a teacher, he had July and August off. One day, hurrying to leave the shop, I had a gooseneck trailer hooked up to our new orange half-ton Dodge truck. I opened the end gate to unlatch the coupler and drive ahead, so the hitch was unlocked before shutting the end gate. Having forgotten that the hitch was not out of the box, I closed the end gate, jumped in the truck, and said to Jerry, "Let's go!" As soon as we started to drive, we heard a loud bang. The end gate curled like a horseshoe, and Jerry sat there laughing while I cussed myself for damaging my new truck, with little money to fix it. This little orange Dodge had an automatic transmission. At the time, everyone thought a standard transmission was needed to tow trailers of this weight.

Well, that truck was the little engine that could. It clocked over 350,000 kilometres before I lost track of its owner. Through the years, it would pull loads of three stacked goosenecks, porches, and twenty-six-foot gooseneck campers through the mountains to British Columbia many times. I don't remember any major repairs. It was a tremendous first company-owned truck—and a Dodge!

"Depending on the trailer model and length, the manufacturing line at Brewster Trailer Manufacturing comprises of three to four assembly stations," I said. We soon found out that the unsold finished product takes up a lot of outside storage, so secure and safe storage was critical. With no outside storage, the finished units had to be shipped directly to a dealer or hauled back to the welding operation for storage.

"The big challenge," I explained to a potential customer, "is to be able to remove the completed, over-sized unit out of the shop and around the ninety-degree corner. After all, there is only forty feet out front before the darn chain link fence blocks their passage from the building. Enter the fine art of operating the forklift towing the unit," I said, noticing that I had their full attention.

"It's possible to safely navigate an oversized unit through a limited opening, to the street for pickup, with precise rotations of ninety degrees. I have to say," I added, "I enjoyed the challenge and did well, especially when it came to the more significant twelve-feet-wide and thirty-feet-long units.

"Watching me one day, my friend Rick laughed and said that stacking and wielding bales in my teen years must have given me an edge. I think I still heard the order, 'Get it done, Doug!' Gradually, we built even larger trailer porches and portable offices, but the majority were sixteen to twenty feet long and ten feet wide. Still, a porch for a mobile home can be as small as six by eight feet, which we also designed and built in components, shipping to dealers and customers for assembly. The mobile homes of that time were twelve to fourteen feet wide and fifty-five to eighty feet long, so a fourteen by forty-foot porch attachment added significant additional space."

Accrued knowledge of detail was imperative in design, especially in our attempt to build a brand-new prototype fifth wheel camper. I'm happy to say it was one of the original units on the market. I brought in an excellent designer, John, who was a past Commodore employee. He assisted in originating the first twenty-fix-foot long prototype camper with a bed installed above the fifth wheel hitch. Our objective was to build the best unit at an affordable price.

Our new concept was much better received than we anticipated. Who knew, many years later, the fifth wheel campers would be as common as straight hitch campers. Dealers were excited. Rick and I hit the road to drive out to the coast of British Columbia to demonstrate the new prototype. I have to say, though, that although it was well-built, the front-boxed design may not have been as appealing as our later, more streamlined design. The smaller fifth wheel units had one axle and were about twenty-two feet long, including six feet over the neck of the truck bed. We hit the road, as enthusiastic as can be, despite the ice and snow. Luckily, we packed truck tire chains. Back at the shop with our staff, I shared the story about the trip over the mountains with a trailer twice the weight of the truck over a morning cup of coffee. "Treacherous," one of the workers said with his lips compressed. Even with those tire chains installed, it was slip and slide on that death-defying trip. But we made it through the Rocky Mountains to Vancouver Island and the Pacific Ocean. The beautiful island, with its warmer winters and numerous trailer dealerships, became a hot spot for our products. Even Darcy and our cousin, Spence, with our forty-foot trailer and three-ton truck, helped transport product through the Rockies, taking the two-hour ferry ride to the island and north to places like Ladysmith and Nanaimo.

—

"I can hardly believe it, Bert," I said. "We're still discussing planning a new fifth wheel trailer design?" Bert was my friend and short-term partner, a successful houseboat builder, operator, and businessman from Sicamous, British Columbia.

"One of our primary design goals was to decrease resistance and provide sufficient headroom for the bed area above the hitch neck," Bert replied. "It had to clear the cargo box of the truck, right, Doug?"

"Yes, I said. "Our truck-over design allows for more room on the main floor, and the rest of the structure is like any travel trailer, featuring a kitchen, bathroom, table, and sleeping accommodation."

Bert smiled, still so pleased. "Yes! And the advantage to a gooseneck hitch is that the higher hitch weight offers a much-improved control of side sway. The winter trip on the icy mountain roads left no doubt in your mind, eh, Doug?"

"Oh no, we understood, upfront and personal, the value of hoisting the trailer's weight over the rear axle of the pulling unit."

"Doug, we could have competed with larger travel trailer manufacturers, no?" he asked, his voice trailing off.

"It's okay, Bert. At that time, we could not compete with the big guys. That is a fact. And remember, too, our camper market developed in conjunction with the 1970s pleasure truck industry, which offered big luxury items."

"Yes," he agreed, sinking into a visitor chair at last. "Increased engine power and large, deep reclining seats for comfy roadside naps."

"I've no regrets, Bert. We enjoyed our adventure!"

After Bert's welcome visit, I reflected upon the downsizing and eventual closure of our trailer manufacturing business. In the late 1970s and early 1980s, I thought that with the higher fuel prices and trailer industry changes it would be difficult for us to compete. Our development and construction business was extremely busy, consuming most of my time.

Talking construction, I explained to Bert that the city of Calgary was expanding Foothills Industrial Park, offering incentives for small- and medium-sized businesses owners to purchase land to build their own buildings. It was a good opportunity, so we decided to amalgamate both our operations and enjoy the additional yard space we so desperately needed at the time. I picked a lot just over an acre in size in the northeast Foothills area. In the middle of the prairies and on the outer edge of Calgary, there were only a couple of other buildings around.

Next, I hired a contractor to level the land, and on weekends I would rent his equipment to load and haul dirt on the site to fill low areas. Perhaps I should have hired contractors to do the job and spent the weekends resting or visiting friends. I enjoyed working with the equipment. Sometimes a change is as good as a rest.

I worked as the general contractor, building our facility by using sub-trades on a concrete block building forty feet by eighty feet, and eighteen feet high. I gave it a flat roof with bright orange metal siding. Orange was our company colour. We built five offices in the front of the new building, which included a meeting room and a large reception area. I divided the main work area in two section: one part for wood production, separate from the spark-flying welding production, and one part for the production assembly of both lines. For fire safety, I made sure the painting was done outside in a portable building that we also built. Given my relative inexperience, I was proud of this accomplishment.

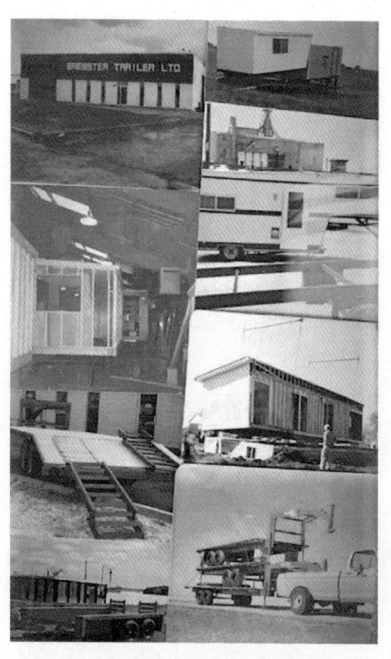

Manufacturing Operation

On average, twenty-five employees operated the production line and the management of the operation. The staff included managers, Rick and Rob; Barb, our office assistant; and excellent drafters and designers. For a couple of years, I was grateful to have my brother Darcy on the welding production line during the winter months, including night shifts. Just as Dad's brother, Calvin, joined him in Peterborough in 1942, so, too, did my "little" brother join me. We also built a couple of structural steel buildings that we erected off-site. Altogether, it was one of the best teams the company ever had, contributing tremendously to the Alberta industry of transportation for freight and equipment.

Our final products, such as, trailers, mobile home porches, portable construction offices, and fifth wheel campers, were shipped mainly west of the Saskatchewan/Manitoba border to Vancouver. With eyes open wide, we always asked two critical business questions:

1. What is missing?
2. What is needed and affordable?

By answering those two fundamental questions, we grew our production to include freight trailers for the cattle and the grain industry, which we proudly showcased at the November Agribition and the Spring Progress Show in Regina, along with many other western agriculture shows and exhibitions.

Inventory financing was not yet readily available during the 1970s as is today, which made it difficult to display our product on a dealer's lot unless we agreed to work on consignment. It meant added risks and inventory financing was our responsibility. Although the operation had become self-supporting, with twenty-five salaried employees and other operating costs, money was tight for our young company, and little extra money available for off-site inventory. GST, our present favourite Federal tax, did not exist until 1991; however, there was a manufacturing government tax that needed to be paid on all units sold. The Alberta government had an excellent program where they helped finance our finished product by buying our completed units until sold.

With credit not available yet, businesses like ours had a tough time showcasing our products for sale over large areas. So, to determine how best to sell and service our products, we had to consider opening outlets in Saskatchewan and British Columbia.

The new manufacturing plant ran at full capacity, including night shifts, mainly supplying Calgary and surrounding area.

Product was also selling well around the Regina area, due to my contacts in my hometown of Earl Grey and in the Craven area, where my family farmed with little competition. Rick's father, Wes, was selling our gooseneck trailers like hotcakes in Red Deer, where he lived and worked, with virtually no competition there as well. The mobile home dealers across the prairies had no trouble selling our Brewster porches.

When a Texaco service station in Lumsden, Saskatchewan came up for sale in 1973, we snapped it up, sub-leased out the restaurant, and turned the location into a Brewster Sales and Service Ltd. outlet, continuing with the gas service. The shop was converted into a trailer service and repair outlet.

—

"None of this would have happened if we'd have just let things be or were content to enjoy the ride," I explain to the class of business students I was invited to address. "In business, evolution is on-going. We set-up the Saskatchewan company called Brewster Sales and Service Ltd., and Harold, a local businessman, helped to develop the operation and efficiently manage it full time for quite a while. With great enthusiasm and with lots of help from Rick, back in Calgary, and all the local staff, including Ken, Art, and Donna, we succeeded in another venture!

Two local women, June and Betty, operated the restaurant, which was part of the building, so none of us had to go far for a delicious home-cooked meal or to gas up. The two service and repair bays were kept busy with servicing product. Initially, we only sold Calgary product—mainly gooseneck trailers for the agriculture community—but before long we were building and selling ski doo trailers and much needed slide-in livestock racks. In addition, we installed hitch installations for trailer owners, including wiring for both gooseneck and straight hitches. Soon enough, we began selling Sno-jet skidoos and McKee snow blowers! Both items were relatively new to consumers, as well. Maybe the brutal Saskatchewan winters and closed, isolated roads and yards contributed to the popularity of these items.

We first began selling snow blowers that attached to the back of a farmer's tractor. A fan backing up a horizontal eight-foot cross auger, which efficiently blew the snow far into the ditch, cleared the road. However, the tractor driver had to continuously look back, which was extremely uncomfortable, especially with a lot of snow. To remove the snow even faster, I designed and manufactured a hitch-and-drive system to attach the snow blower to the front of a tractor. With the support of my excellent staff, I designed the gears and chain drive for the snow blower to change direction when attached to the front of the tractor. However, Harold called me one Saturday morning to tell me that a farmer was trying to clear his badly snowed-in road, but the brand-new snow blower he had just purchased from us was rotating backward, not forward. Imagine. By Sunday morning we had removed and re-welded the chains and gears of his new machine, and by late afternoon, the farmer's road was clear of snow.

Upon reflection, I consider that I never laid on my bed as a kid and dreamed I would do design work. Not once. Instead, as an adult, I responded to the real needs, and opportunities, that cropped up.

—

One day, a company out of Winnipeg called Canadian Tool and Die, who manufactured ski doo trailers in addition to their central axle and hub business, decided to get out of the trailer business. What a sweet surprise! Their stock and jigs were up for sale. Harold made the trip to Winnipeg, and when our offer was accepted, he brought back three semi-trailer loads of trailer parts and building jigs to our Lumsden location. Even after building as many trailers as we already had, we still had enough inventory of steel pipe, couplers, and hitches to last us years, both in Lumsden and Calgary.

Once Brewster Sales and Service Ltd. was well-established, we continued the search for a market outlet around Vancouver. Rick and Margo decided to move to British Columbia, so I purchased the Calgary home they vacated. It was a cozy place where I had often found warm refuge while wading through my prolonged separation. But it was hard to see my dear friends go.

—

After being away from the acreage in Airdrie for a while, I drove out to check on it. When I got there, I discovered that my young horse, which I was trying to break for riding, had broken out of the pasture. Tramping around the grassy fields, I found her at a nearby acreage enjoying the attention of three children. Their livestock looked superbly well-cared for, so I called and asked their dad if I could give her to the kids. He gladly agreed and the happy munchkins thanked me enthusiastically. I drove away thinking, *no more horses for me*, but little did I know what the future had in store.

That month, I reluctantly sold the acreage with the unfinished little cottage I had built overlooking the Rocky Mountains, near Airdrie. I also sold the mobile home I purchased from Commodore while working there.

—

"It's me again, Doug. Harvey. I've got a new idea."
"Another stunt pilot?" I joked.

"No, Doug. It is time we set up a limited partnership with a third partner to develop commercial properties. Terragreen Properties will do the marketing, and I'll manage the business, get the financing, and develop new projects," said Harvey.

"And I'll do the general contracting with my new incorporated company, Douglas Industries Ltd.," I added.

"Exactly what I hoped you'd say, Doug."

"I've already been working on a second project, Harvey. Since various cities, including Calgary, are offering low-priced industrial land for business development in exchange for future property tax revenue, I have already taken an option on a large commercial lot in Regina, located in an industrial area. It has to be developed within two years though, Harvey."

"Done."

We named our development company TOGO after a town in eastern Saskatchewan. Three proposals were actualized that day:

1. I transferred my Regina industrial property into the limited partnership.
2. Terragreen contributed land for a fourteen-unit condo in High River, Alberta.
3. Harvey obtained the working capital for the operation.

Harvey and I also set up another company named Caria Properties with a commercial building project close to my original Brewster Trailer Manufacturing building. In my wisdom, or lack thereof, I committed to leasing a good part of this new building for our trailer manufacturing business in order to secure the land from the city of Calgary. I was also facing the raw fact that the market for our freight trailers was saturated with new industry manufacturers. With gas prices on the rise, it seemed that the demand for our economy gooseneck trailers was fast disappearing. I felt the strain. My interest was in high production standard trailers and not the custom-made cargo trailers customers now wanted. Here it was, October 1976, and it was the beginning of the end for the Brewster gooseneck and straight hitch trailers. Amazingly, ten years later I was proven wrong with mass trailer production taking hold, with consumer credit and large sales lots with massive inventory of trailers.

My personal life in the 1970s was one of independence. I was comfortable on my own. With good friends and a caring family, I was content. I was enjoying my life and reluctant to engage in another personal relationship, since my last one was fraught with tension. However, things were about to change for me.

It began with Lloyd, who had worked for me at Commodore and had been keeping in touch. Jenny, his wife, was a great gal and worked for Shell Canada in downtown Calgary. They thought I should *really* meet one of her friends. I brushed it off and continued with my independent life, as I had got to know it over the last couple of years. As I was relaxing one evening, Lloyd called me to ask about meeting for a drink when they returned for a skiing trip in the mountains. Upon meeting them, I was introduced to Jennie's friend Donna, who she worked with at Shell. We all had a drink, and I learned that Donna was a city girl, a bit younger than me, and had lived in Calgary with her family for most of her life. She also loved sports. Although I thought she was attractive, I felt we were from different worlds.

On my walk home after meeting her, I thought, *if only things were different, and I had met her four years before when I was a professional, working, and playing in the big city life.* A few weeks later, Lloyd and Jennie had a house party we were both invited to attend. Well, I got quite a surprise when I saw Donna. She was all dressed up, no longer wearing her ski clothes after a day on the slopes. I guess I forgot all about my short comings because I decided to pursue her. I do not think the feeling was mutual—at least not at the start—between this city girl and this unpolished, poorly dressed, not-so-sharp guy, but I was persistent.

Donna and Jenny were on the Shell softball team. With an invitation, I began to make time outside of work to attend games and socialize with them afterward. Donna was independent. She had her own car, so we would both go our own way after these get togethers. Later that summer during a ball game, Donna injured her knee and could not drive, so I offered to drive her home, which she accepted. According to her, I said something ridiculous like, "Maybe we could go dancing sometime." What may have helped me in my quest was that, unfortunately for her, Donna wound up with a leg cast, and I was there—Johnny on the spot—to take her to the hot springs in Banff to heal.

—

Early in our relationship, Lloyd and Jenny asked Donna and I to go camping in their motor home in Windermere British Columbia for a weekend. I had a great time visiting and relaxing; however, the plan had been that my sister, Donna, and her husband, Bill, were going to fly to Calgary, borrow my car, and come to visit. But we did not get to see them. Although we had given them the campground name, I guess we forgot to give them Lloyd's last name or the type of motor home we were in. I am sure there was some suspicion about us trying to disappear on purpose, as they'd not yet met Donna.

Another time, Donna invited me to join her and some friends at the Calgary Stampede. While reasonably up to date with business attire, my closet was otherwise bare of casual wear. I guess it was apparent that I had gone and purchased a new wardrobe for the occasion, which included stiff new cowboy boots. I felt so lucky to be going out with Donna, a bright and beautiful big-city girl that I endured the pain of those new boots.

The fashion of the 1970s was mini-skirts and bell-bottom pants. On a work trip to England, Donna took some time to enjoy herself and learned how to wear miniskirts. She dressed beautifully.

Soon, Donna accompanied me on a business trip to Los Angeles, where we squeezed in a weekend at Disneyland and Knott's Berry Farm, an amusement park with farm animals where piglets had recently been born. Donna was so intrigued and excited; she had no idea that one day we would be raising and selling our own piglets.

Donna and I continued seeing each other and visiting her family's campground site in Windermere British Columbia. When Donna was there, I'd drive to the lake to meet her. She'd always anticipate my exact arrival time and walk out to meet me. There were no cell phones in 1974, and I was always notoriously late, but for a short time I became super punctual!

Donna's mother, Mildred, began inviting me to supper. Feeling her sincere welcome and being that the invitation enabled me to see Donna after work, I was soon a regular supper guest. Her parents were genuine people, and the meals were so enjoyable. If they had an ulterior motive, I sure took the bait.

One weekend at Brewster Trailer Manufacturing's Calgary's plant, Donna was there with the staff and wielding a staple gun to secure the wall panelling to finish building a rush order of eight foot by eight-foot knockdown porches. That order most definitely went out on time. On another weekend, in the winter, we flew to Regina to visit Brewster Sales and Service Ltd. in Lumsden, Saskatchewan. Harold had arranged for us to ski doo to their farm, fifteen miles north out of the Qu'Appelle Valley. The business review finished at about 9:00 p.m., and despite the cold night, Donna and I set out on a Sno-Jet skidoo. She hung on behind me, not realizing my lack of experience. We drove in the ditches along the highway to Craven, and then six miles north to the farm. How we made it through the huge snowbanks we encountered is a mystery, especially considering my limited experience snowmobile driving. Harold and Audrey insisted we stay the night and made us feel comfortable in their home. This was Donna's first night on a farm and the beginning of our Saskatchewan visits and stays, which became a big part of our lives together. Harold and Audrey became our good friends, and they remain so to this day.

I soon learned of Donna's expertise. She could handle herself with everyone she met, she loved learning, and she was happy to share her sparkling opinions and ideas with others. Later in our relationship, she damaged my car by hitting a deer. She told me afterward that my forgiving attitude about the crash sealed the deal for her. Thanks to Lloyd and Jenny, I met and fell in love with Donna, the woman who understands and loves me, as I do her.

—

I had been living in the basement of the house I had purchased from Rick and Margo when they moved to British Columbia, but I later purchased a house in Ogden, Calgary, closer to work. Around the same time, Donna decided it was time for her to get an apartment of her own. London House was only three blocks from the Shell office, and it had an apartment on the twenty-sixth floor with a balcony and heated underground parking. Donna practiced her cooking skills with enthusiasm, and I knew to say nothing at all about missing her mother's delicious meals. I soon began to involve Donna in my businesses, tapping into her skills to help.

—

On our travels, we observed a neat metal chain plant hanger that we started to make after hours. We had boxes of parts with jigs and various tools to bend and crimp metal strips and chain in our apartment. Once painted and wrapped, I would distribute them to various variety stores in downtown Calgary for sale.

As you might expect, I eventually rented out my house in Ogden. Little did I know, though, that I had rented my place to a con artist. He failed to add his name to the utilities and phone service. When huge bills arrived in my name, I discovered I was one among many who had been duped by the so-called "antique dealer." At the time, I did not make a huge deal about it; I was happy with Donna.

In passing through Las Vegas, Nevada on a Commodore business trip, I was intrigued by the luxury and glamour—money seeming to be no object. I knew that one day I would like to return, so I suggested to Donna we take a weekend trip to there. Even though she had been there after winning a free trip a few months earlier, she was excited about returning and showing me the strip and Fremont street. Soon we were on our trip, enjoying the sights and spending our hard-earned money in the casinos. I had grown a mustache to help me look older when here for the Commodore job, but Donna was not very keen about it, so I surprised her by shaving it off. I also surprised her with a pearl promise ring. She was very shocked and excited, we both were. We could not wait to return to Calgary and let everyone know. Her family and our Calgary friends were elated on our return. We also had much support from my family and employees in my business.

Life was great. Never in my wildest dreams would I have guessed I would be so happy and content about the future. Lucky me.

Spring 1975, Donna and I became engaged. I had picked out my first diamond ring. As I recall, knowing how important things like this are to this city girl, I picked out a ring with a raised diamond. I made sure it was her style and that she would love the ring. Unfortunately for Donna, I am not one of those on-your-knee or ask-her-father type of guys. Nevertheless, our engagement was a meaningful moment, with us looking out over Calgary from our twenty-sixth floor apartment balcony. Donna loved the raised diamond ring.

We worked as a team early in our relationship, respecting each other's views and needs. It is uncanny and unbelievable how even years later how we both get the same idea or thought at the exact same time, whether it be our early meetings on the Windermere bridge, calling each other miles apart, or bringing up the same topic as the other person. Could this be true love?

The summer went by quickly, and Donna was soon busy going to her bridal showers. We attended many family, friends, and employee parties in Calgary and Saskatchewan. Donna, with the help from family, did most of the planning for our big day while I was playing the busy working guy.

There was one thing in our relationship that needed addressing though. Donna and her family were Catholic, whereas I had been brought up in the United Church and had been divorced. So we met with her family's priest to discuss our options. I was open to ways of dealing with the problem, knowing it was important for Donna. However, the Catholic Church had strict rules about marriage in the 1970s. My prior marriage needed to be annulled. The procedure of showing my marriage had not existed meant involving all parties, which seemed impossible at the time. We had no communication with Merle, lawyers were involved, and there was the issue of my legal rights to see Bettyanne. I felt terrible for putting Donna in this situation, and for a guy who prides himself on problem solving, I did not have a solution. Looking to me, I reassured her that we could do this and have our marriage ceremony in a United Church. To this day, I still have a sore spot for the Catholic Church and the decision Donna was forced to make.

On December 27, 1975, Donna and I married in a Calgary United Church, in a small, elegant ceremony, with our immediate family in attendance. Donna looked beautiful in her high-necked, full-length gown. Our reception was at the Calgary Inn on a bright, sunny afternoon. It was only two blocks away from our apartment.

We left immediately on a short honeymoon to the natural wonderland of Lake Louise and the Rocky Mountains. The stunning view of the chateau and lake was mystical. It was a perfect spot to start our life together. That place continues to be a favourite location for us to celebrate anniversaries and spend weekends. Over the years, I have spent many hours relaxing in thought there—sometimes writing notes on my memoir or earlier times, sometimes contemplating a mediation meeting or design issue. The main floor lounge and afternoon sitting area, with its large windows and fine woodwork, look over beautiful Lake Louise, providing me with tranquility and peace of mind.

Upon returning to Calgary from the Rocky Mountains, we flew to Maui, a Hawaiian island in the Pacific Ocean, halfway across the ocean to China. We went for some time away from winter weather and our hectic everyday activities. Although I usually worked on projects and needed to be in touch with the staff back in Calgary and Lumsden during the day, Donna enjoyed the beach and sunshine, reading her book and trying to get a great tan to impress everyone at home. We did a lot of site seeing and enjoyed the warm evenings and entertainment. We happened to meet my past accounting clients, Peter and Wilma and their family. On hearing of our marriage, they graciously sent a bottle of chilled champagne to our suite—a thoughtfulness I have always remembered.

Upon our return from Maui, in her exquisite yellow Hawaiian gown and glowing tan, Donna and I distributed two hundred fresh Hawaiian orchids we had brought back on the plane. We brought them for her sister Janet and Scott's Calgary wedding on January 31, 1976 which included a joint reception with friends and family. Mildred, Donna's mother, wished us all the joy she had known in her married life with Alex, Donna's father, who were also celebrating their wedding anniversary. What a wonderful time it was.

—

The summer after our winter snowmobile ride to Harold and Audrey's farm, we became good friends with Dick and his wife, Mick, at Silton—a little village only five miles west of Harold's farm. Dick operated a service station in Silton and asked us to join them in buying a couple of lots close by in the Qu'Appelle Valley, next to Last Mountain Lake. We loved the lot and its view of the lake. We purchased it with a plan to one day build a twelve by forty-foot cottage at Brewster Trailer Manufacturing with my sister and her family back in Calgary, then transporting it to our new lot in Saskatchewan. We were excited.

That first summer of mine and Donna's marriage, while waiting for our Calgary house to be built, I needed to spend time at the Lumsden operation. Donna and I pitched my old green canvas tent on our treeless lot, overlooking the lake, and we lived there for the summer. Donna still has not forgiven me for supposedly abandoning her on those hot, summer Saskatchewan days. I'm sure it was not as bad as she now let's on, with the beach having been two blocks away and my sister visiting. We had to be up and out of the tent by 6:00 a.m. most days because of the heat. I reasoned that Donna loved the beach and tanning, so I would go off to the office and the conveniences of Lumsden, Saskatchewan. Mom and Dad felt sorry for Donna, so isolated, so they loaned her a car so that she could visit and be mobile.

The next year, our finished cottage stood on the lot next to Dick's double-wide home, also produced by Brewster Trailer

Manufacturing in Calgary. We spent many happy times with family and friends, swimming and boating with the sixteen-footer called DeeDee, which Donna and I purchased soon after we met. Water skiing was great fun; our boat had a large enough outboard motor that we could pull up to three skiers at once, which challenged both the driver and the skiers. The boat had what they called a camper cover, which resembled a small cabin cruiser with windshield wipers. I enjoyed this feature especially when fishing and boating on rainy and windy days.

On my birthday in 1977, we found some coal from an old railway coal shed for our wood burning fireplace and stayed in our Saskatchewan Cottage, that cold January weekend. A mystical moon shone over the lake and entered our patio door; it was a magical moment of beauty. Thank goodness the January Saskatchewan weather cooperated with us city slickers at our little summer cottage in Saskatchewan Beach.

—

During the next couple of years, the businesses in both Alberta and Saskatchewan kept me extraordinarily busy. Describing the intensity is hard. Brewster Trailer and Douglas Industries Ltd. built and erected a steel building for a company named Ronaco, owned by Ernie, my fellow chartered accountant and fishing and camping buddy. Dick, our partner at Silton, Saskatchewan, decided he was ready to retire and shut down our little porch and office manufacturing company named Brewster Sales & Service Ltd. We had set up this business in Silton a few years before, when we were neighbours at Saskatchewan Beach, after Dick demonstrated his interested in Brewster Trailer.

Two brothers I had grown up with back in Earl Grey, Bob and Johnny, built many units for Brewster Sales and Service Ltd. in that little Silton shop, and Darcy supplied the steel frames he built in Dad's farm shop. I remember Bob, who was a couple of years older than me, back when he worked for my dad driving tractor in 1952. One time he backed our Massey-Harris 44 tractor up, accidentally hooking our International 9 one-way tiller on top of it with me riding beside him. I guess his foot slipped off the clutch, but he made me swear to secrecy. Our friend and partner Dick enjoyed being on the road, and Donna and I were never surprised when he and Mick would show up in Calgary to bring us up to date on the business. We all toasted the success of our venture, and we wished them well.

—

In 1977, I discover that the prime Red Deer commercial property I had purchased was ready for development. Douglas Industries Ltd. erected and finished the metal building with the steel structure, built by Brewster Trailer Manufacturing in the Calgary plant and hauled to Red Deer. I had to hire a big cat dozer to clear the bush off the property and prepare it for the building and parking site. Little did I know I'd clear many acres of bushland in the future with a cat dozer.

Vern (a good friend and employee from Douglas Industries Ltd), his crew, and I would leave Calgary at 6.30 a.m. that winter to log in as many daylight building hours as possible before heading on the one and a half hour drive home on Highway #2. Even though our old motor home sat on a site with multiple electric heaters running, it was freezing cold—. Running our drills and handling metal was an endurance contest. At night, the generators and tools were returned to the heated rental company, so they would be warm the next morning rather than frozen solid. Our dinner breaks at the local heated fast food outlet were a huge relief. Gratefully, on returning with his family from British Columbia, Rick helped finish the roof, and soon the property was leased to a new truck franchise. We returned to Green Acres with the job done and no idea of Alberta's impending 1982 economic crash, which would result in the building coming back to our ownership.

Over this five-year development period, we completed a commercial building, a fifteen-hundred square foot residence in Red Deer, and a sixteen-unit condo complex in High River. Phase One of the commercial building project was in Regina, along with three commercial buildings in Calgary and our three thousand square foot house in Pump Hill. I was consumed by the contracting industry and by my passionate preoccupation with the Alberta business environment. Engrossed by the opportunities I saw all around me, my time monopolized and I became wrapped up in my relentless job requirements. The wheel no longer spun. It whirled.

I accused myself of working too hard and wondered when I started piling so much on in an almost compulsive manner. I saw folks all around me doing a good day's work and heading home to their families for supper and a pleasant evening with loved ones. Why was...pinning top among a far more routine-oriented population who knew when it was time to go home? My answer: My work was engaging. I liked my work; I was never bored, and one thing led to another. Also, I relied on my competent partner, my wife, Donna, to keep me informed. There was no forgetting on my part, with Donna at the helm. I was grateful for that, because I loved my family and wanted my business talents to enhance our lives, not deprive them.

I was fortunate to have Donna as my partner. Donna had worked her way up the chain of command in the competitive oil industry and thoroughly understood management, its responsibility, and its sacrifices. Her personal and organizational skills helped tremendously with the success and happiness of our family, extended family, and dear friends.

—

Rob, a fellow who had worked for me at the Commodore manufacturing operation in Calgary, came to work for us in 1972 when Brewster Trailer Manufacturing was expanding in a Calgary building once used for growing mushrooms. As our success grew, and thereby Rob's success grew, he became part of the team. I relied heavily on him and, at times on Rick and Dorothy as well. Rob became a trusted employee with a salary, company vehicle, and travel expenses. Unbeknown to me at the time, he was taking a bit more of our profits by stealing and personally selling inventory and equipment from the plant, as well as taking supplier kickbacks. While I knew that we were missing product, it appeared to me to be outsiders. I had no indication it was any of my other trusted employees.

Another past Commodore employee, Larry Graham, stopped by one day because he was interested in buying product. Larry was multi-talented and a risk taker with a lot of similarities to me. He would go on to become one of my good friends. Larry reminded me of the day at Commodore when he had nearly up and quit, in 1969. Apparently, unbeknownst to me, he had told the crew that the parking lot needed serious de-icing, but no one moved. Larry repeated himself and said again, "The area needs a good scraping before we attempt to move the two-ton crane."

"Nah! It's fine," one staffer told him.

Another worker piped up and said, "The crane's so heavy it'll crack the ice as soon as it starts up. It'll grind it like sugar."

Sure enough, the minute Larry turned on the ignition and clicked off the brakes, away the crane slide across the lot, slamming into one of the company's brand-new mobile homes, fresh off the assembly line.

"Braking made not a whit of difference. Away it careened," said Larry. He swallowed half an egg salad sandwich and continued. "I turned it off, jumped out, and slipped and slid my way back to the office. I slammed the keys on the front counter and barked, 'I quit! You fellas say the lot is good? It nearly destroyed me and the expensive machine! Don't think I'm paying the damages!'

"No one said a word as Doug, our young chartered accountant in charge approached me. He slid the keys back over the counter to me. 'Don't resign over that, Larry,' he said, loud and clear.

'It's us who put you in danger out there. Sorry about that.'"

As Larry finished up his story, sealing our friendship, Barb handed him a frosted donut. "Melted icing sugar," she smiled, and we all laughed.

—

In 1977, I decided our Lumsden axle operation needed some personnel changes and moved Rob to manage the operation. I helped him buy a home in Lumsden for him and his wife, with all moving expenses paid. At the time, I felt Rob had deserved the promotion and had no concerns about his character. If only I could have seen the future, my decision making would have been much different.

Our Calgary operation had had items go missing in the past; however, even with the authorities involved, there were no signs of employee theft. But I would soon learn that Rob had been removing material and equipment for some time, and I suspect he felt he deserved it. Did he feel guilty or ashamed? I don't think so. In truth, it was he who felt robbed. He likely thought that for all his hard work, doing more than his share for the company, he was entitled to higher wages and profits.

—

"A quotation by Honouree de Balzac, Doug," said Mr. Code, the vice principal and my favourite high school teacher at the old four-room Earl Grey School. I liked the slow-speaking Mr. Code. I liked the way he listened and the way he cared about me. Whether I biked or walked the last half mile to school, or after stowing Dad's truck off on a side road near town, I knew Mr. Code would be there.

"It's just a short line: 'A fast flow of words is a sure sign of duplicity.' Maybe it sounds better in French, but it's about smooth talkers who fool you. It's about unsavory characters who hide in plain sight," I answered him.

—

Funny how that that brief exchange from 1958 came back to me exactly twenty years later when I discovered our smooth-talking Rob was busy stealing from the axle operation. I'd trusted him. I had honoured him with the top managerial position of the trailer axle operation, perhaps because he was so successful at convincing me of his qualification.

"During our rounds, Mr. Brewster, we noticed an increasing number of crates in a field outside Regina," said the RCMP officer. "We got curious about the stockpiling in an unauthorized zone. Our crew traced the material back to you," said Sergeant Wright.

"I'm shocked!" I replied. How long had Rob been stealing from us? And worse, how much had he stolen? I felt like I'd been blindfolded and beaten up with my hands tied behind my back.

In the days following, I carried a cloying heaviness with me, as though I was weighted down with sandbags of wet ash. It happened like this:

"I smell a rat," said my brother-in-law on that late fall day in 1978. Bill was helping me hang interior doors at my new four-plex project in Lumsden when we began inhaling acrid smoke, each breath growing grimier than the last.

My axle plant! I thought. The two of us dashed the three blocks to view my building in flames. Screeching to a breathless halt at the plant, the one remaining fire-scorched machine steamed smoke outside the destroyed building.

"The forklifts out, Doug, but not much else. It is horrible. Oh my gosh… your operation is destroyed. I am gutted. Look at this, charcoal chunks of dust," said Art, a caring employee now unemployed.

Before I could compute the ruin, the phone started ringing. It was legal calls.

"Where were you, exactly, when the fire started, Mr. Brewster?" an insurance inspector asked, a hint of suspicion in his voice. "Exactly how far?" he prodded. "Don't know if your coverage is adequate, either."

"I'm speechless, Donna," I said, calling Calgary to tell her of the fire. My wife did not try to dissuade me of what she, too, felt. Was this man, a fellow we chatted and joked with every week, an arsonist? Rob had looked into our eyes. We had been utterly fooled by a duplicitous expert, a practiced deceiver whom I'd fully trusted. I felt my hands and fingers swell with shock.

Our young employee, Art, had the presence of mind to save our forklift. In consoling me, Donna said "Even though Art risked injury to himself, he drove it through the large overhead door," Yes, he was a good guy and a man of integrity.

Once charged with theft by the RCMP, Rob must have decided to set the fire to cover-up obvious inventory shortages. It was interesting that his friend, a fireman from Calgary, visited Lumsden just before the fire.

"It's one thing to be targeted by strangers," said Sergeant Wright during our last visit, "but it is worse when it's on the inside, from folk you hand Christmas turkeys to and toast at New Year's."

Our recovery took time. Rob was charged and sentenced. It mattered not to me; I simply left it with the police.

—

Hi, Doug, Harvey here."

"Good morning, my friend. Have you decided?

"Yes, we have," he said that morning in February 1979. "Look, Doug, you oppose our new plan. You object to TOGO taking over the general contracting for which your company, Douglas Industries Ltd., is presently responsible. Is that correct?

"Yes, it is, Harvey. I object to that plan."

"As is your right. I guess we will make you an offer then. How about we buy out your share in TOGO?"

"Fine, Harvey. Give me a fair settlement, and I'll move on."

I was disappointed. Our team at Douglas Industries Ltd. were the construction experts, and I believed our expertise was necessary for the future success of TOGO. But as the future would soon reveal, I was again lucky with my decision to leave that joint venture.

—

"Call your mother, Donna!"

"But you said not to, Doug!"

"I know, but I thought I'd be home! You're early!"

Rob, still employed by us at this time, gave me a ride from Brewster Sales and Service in Lumsden to Regina. I hopped a flight home, and I made it to the hospital in the nick of time.

In June 1978, we became the proud parents of our healthy baby boy, Stephen Douglas. Donna recovered well and sailed into loving motherhood. For the first year of his life, Stephen enjoyed his doting Mom, me after work hours, and Bettyanne, who was thrilled to hold her brother during her weekend visits.

Fifteen months later, I once again set out to Brewster Trailer Manufacturing's offices, but this time, when I was about halfway to work, I turned myself around. I knew I should be home with Donna. I was relieved to be beside her for the birth of our baby daughter at the Calgary Foothills Hospital. After three hours of brave labour, Donna gave birth to Catherine Dawn in October 1979. From the outset, Donna and I delighted in the newest member of our family. We soon saw our daughter's spunk and motivation.

—

Pump Hill house

Two years earlier, my company (Douglas Industries Ltd.) along with Vern and his crew, built our dream house. Donna and I spent many hours designing our new house and purchased a treed corner lot in the new and upcoming area called Pump Hill. Our home featured three floors with an attached double garage, open spaces, indoor gas grill, two brick fireplaces, and oak railings. There were three bedrooms on the second floor: one for Bettyanne's weekend and holiday visits, our master bedroom, and a spare room for visitors. With her savings from Shell Canada, Donna bought our beautiful oak dining room suite. She was immensely proud of it, and with the kind help from her family, she did our home decorating.

We were happy enjoying our new home; however, as things turned out, it only lasted for four years, due to the economic downturn and the Balzac land purchase.

THE SLOW DOWN AND BALZAC FARMING

I WAS THIRTY-SIX YEARS OLD, MARRIED FOR four years now, and Stephen six months old. I wanted to slow down and enjoy my family. With no preview of the stunning 1982 Alberta financial crash soon to come, I began to close Brewster Trailer Manufacturing. I sold my Caria and Togo interests and ended my high interest operating loan for the 9th Avenue property. I made no money on the project, but luckily, I escaped unscathed. Soon, in the throes of the energy crash, we saw streets fill with families who had lost their jobs and had to leave their mortgaged homes empty. We had narrowly missed a disaster.

—

Partnering with our great friends, Donna and I collaborated on our next venture: the purchase of a 160-acre hobby farm two miles east of Balzac, Alberta. Rick and Margo, and Bonnie and Bradley, moved into the old farmhouse on the land, and Donna and I stationed our old little motor home close by in a nice sheltered area with a picnic table and newly built fire pit. While we continued living in our Pump Hill house, we were at the farm most weekends and all holidays.

Together, the three of us couples improved the farmyard, added livestock, started a hog operation, and added land and grain storage. In the evenings, we would all trek out to the fields to enjoy magical times with up to seven young children around our laughter-filled campfires, toasting marshmallows, singing, and in winter, skating on our natural ravine ice. Our six years of fun, friendship, and hard work also included our good friends Vern, Ann, and their kids, Shane, Sheila, and Susan. They drove out to Green Acres on weekends to visit and help us. I treasure these memories.

Vern and I even found time to go north of Fort McMurray, Alberta with Scott, my brother-in-law and friend, to a fly-in-fishing camp. Although our partying got a little out of hand, we had a good time with lots of fishing and boating, a favourite pass time of mine.

That first winter, Rick and Margo's water line from the well froze, leaving them without water for themselves or their animals. We took shifts descending the eight-foot-deep well with a pressure washer, attempting to thaw the line under the ground to the house. At last defeated, we laid a new hose on top of the ground during the day, and at night we stored it in their heated porch. Once the frost was gone in the spring, we dug an eight-foot-deep trench with our old backhoe to install a new water line...until the next glitch. When we were nearly finished the job, we came across a huge thirteen foot in diameter bolder, which could not be moved and forced us to reroute the entire line from the well.

Our hobby farm was naturally named Green Acre Farms, after the TV show with Eva Gabor and Eddie Albert called "Green

Acres," due to similarities in our inexperience, equipment breakdowns, and livestock getting into the next farmer's crop, or worse, his wife's garden.

Initially, everyone was involved with the livestock and farming our land. We all had fresh milk, meat, eggs, and garden vegetables. By 1982, I was increasing cultivated acres by renting small parcels of land from acreage owners and foreclosed quarter sections from financial institutions. The initial farm equipment we used was another matter, because as our friend Vern said quite accurately to a farm machinery dealer one day, "We're basically looking for junk."

In the late 1970s, there was a large seasonal farm auction east of Calgary called Teske Auctions. It became quite the routine for Rick, Vern, and I to attend every sale, preview the equipment, and bid. The two days following the auction were taken up moving our prized acquisitions to the farm for assessment and repair. Almost all of our equipment—combine, swather, cultivator, and many other valuable items—came from this legionary auction.

—

Imagine selling Donna and my first beautiful three thousand square foot home in Pump Hill to move into a 720 square foot trailer, with young children! Hello Balzac, Alberta. However, the move to Green Acres became a very memorable part of our family's life.

In 1980, our growing family was Donna's full-time job, but she found time for gardening, yard work, and helping with farm chores. My wife, my business consultant with whom I bounced every idea around with and whose opinions shaped and modified my own—would not only strategize with me about my newest notions, but also monitor and maintain the details of my past ventures.

Our first years living in Balzac took me back to milking a cow, separating the cream, having fresh farm eggs and home-made bread. Donna took a Chinese cooking class, and along with her fudge and chocolate, our kitchen was fun.

Balzac farm

I found time to join the Mount Royal Student Association's Wyckham House Board, where I advised the association on their campus business ventures and money-making opportunities that guaranteed struggling students a smoother ride. I represented the school board at special functions, which I enjoyed doing. On one occasion, at an award ceremony, Donna and I had the pleasure of dining with the Honourable Grant McEwan, Lieutenant Governor of Alberta, and after chatting with him over dinner, we were so honoured and impressed at being able to meet him and hear his great opinions and ideas.

I also spent a couple years on the Rocky View County of Alberta's planning commission. Our committee advised the Rocky View's council on subdivisions and proposed economic development in the large area surrounding Calgary.

Our first business venture on the hobby farm was a small modern weanling hog facility, to raise and sell weanling pigs for farmers to feed and raise for market. A lot of work was involved, and with Vern's help and expertise, it ran for a few years before the price cycle for hogs dropped, resulting in little to no profit. However, we did have fun building and operating the business. We would all sit together on a Saturday night, enjoying each other's company while waiting for a new batch of piglets—their mother might need our help.

One Christmas morning Rick was on chore duty. For fun, Donna placed chunks of coal, with red ribbons from Santa, throughout the barn. Surprise, surprise!

—

In 1980, I purchased a small commercial lot with a house on 9th Avenue in Calgary's southeast. Since the purchase price was attractive, I considered it a good investment for the future. I rented out the house on the lot and soon entered what I've heard termed a landlord's nightmare. At first, the rent was paid in cash by three individuals. A year later, I was tracking down a dozen individuals for their respective shares. I worried about the repeated promises of next week, a constant rotation of renters, and interior damage.

One day, the Calgary police called about a complaint from neighbours. A naked teenager was on my roof, tossing liquor bottles down my chimney. I asked the police for their advice and was told to meet them at the house, where we saw a large group sprawled over the porch and lawn. They suggest I give the tenants their eviction notice immediately.

At the back door, I shouted, "People, you're all out as of right now. Please move your stuff to the front lawn." We helped them remove everything from the house and onto the front yard. With everyone out on the grass, I locked the door and secured the property. Shortly after, Steve, still with Brewster Trailer Manufacturing at our Balzac farm shop, demolished the house. He used the old backhoe that once dug the water trench and the hoist gooseneck trailer that we still have on the farm.

Douglas Industries Ltd. came to life again and found a tenant to lease a small commercial building that we were building on the old house site on 9 Avenue. Unfortunately, just as we finished construction in 1982, the operating loan interest for it topped at twenty-two percent. The signed lease was no longer worth the paper it was written on. Due to the market, the tenant could rent a similar property for much less. A deal was struck between the mortgage lender, the prospective tenant, and us. We broke even on the project and the 9th Avenue property was gone.

In the early 1980's, the interest rate for lending money, hit an all-time high over twenty percent and inflation was rising. Our future becoming clear to us, Donna and I decided to auction our Brewster Trailer Manufacturing's assets, our mobile home, and the farm equipment we no longer needed. We called our good friend and talented auctioneer, Larry, and asked for his help. A sale was arranged for June 12, 1983. It was an emotional day filled with hard work for me, my family, and my close friends. I participated with mixed feelings, seeing my company's assets disappear down the road with their new owners. The farm looked bare; I had not realized the amount of equipment left from our manufacturing businesses. But we looked ahead with hope for the future.

—

In 1982, we began raising cattle for sale at Green Acre Farms. Our hobby farm was comprised of horses, chickens, pigs, and two dairy calves to raise for our meat—as every good farmer would do. We purchased five yearlings—Hereford heifers that Vern trailered with his trusty jeep from his father in-law's farm north of Moose Jaw.

When Vern arrived late in the day and with no corrals, we unloaded the animals directly into the fenced pasture with our Holstein dairy calves. The trail gate was wide open, and the newly arrived yearlings got spooked and tore off in every direction, through the fence and out of the pasture. We searched until dark in neighbours' fields of ravens and bushes where the scared animals may have hidden, but no luck. Exhausted, we quietly agreed to search again in the morning.

In the early morning light, a retired cowboy from the area galloped into our yard with large loops of hemp rope hanging from his horse's side. Tall in the saddle, he announced he would help catch our cattle. "Thank you! Thank you so much," Margo and the rest of us chorused as we donned our boots to join this kind stranger.

"Always glad to return a favour, Doug," he said with a smile I vaguely recalled. "Remember cramming this six-footer into

Bernie's Vaux Wagon on your birthday twenty odd years ago? It was your first year of U of S. Remember your motel party?"

"Brett! We boarded together at my first rooming house?"

"Yep, the one I got kicked out of if you recall."

"Oh my gosh, Bernie and Jerry got you upstairs to your bed with some obviously unintentional noise, and then I had to go back and clean up that motel room in a hurry."

"Yep. I could not help that night, or hold my drink back then, but you fellows made darn sure I did not freeze to death in a winter ditch. I'm here to return the favour," said Brett.

In the late afternoon, he returned with two lassoed Herefords being pulled behind his horse. Hours later, he located three more of the strays. "Found 'em quivering in fright behind a patch of prairie bulrushes three miles from the farm," said Brett. "They must be hungry as can be, because they didn't resist being led home." Now safe, the heifers hastened to the feed trough. Brett declined our offer to join us for one of Margo's excellent suppers. Instead, saluting us all with a tip to his wide-brimmed beaver blend hat, he galloped off into the evening light, his quality lassoes hanging at the ready at the side of his well-brushed steed.

—

When a farmer up the road lost his home due to the credit union and the financial crises, I was able to rent his land. However, the farmer and his family first needed to complete an exit auction sale. Angry about his loss, the farmer stripped the house of everything, including the carpet. He sabotaged the water-well system and sold all the fixed cattle corral and fencing. I was reimbursed for the repairs and able to rent the home and run the land. It was such a shame to see, firsthand, the results of poor planning, poor timing, and personal suffering.

In 1983, I decided to add a herd of purebred Limousin cattle to our operation. Cattle sheds, corrals, and handling facilities were added and converted on our farm. Other than our new farmhouse, the yard comprised of repaired old buildings and corrals built from used lumber and not painted. We were proud of and content in our work.

My sister and her family joined us in a partnership on some of the cattle purchased from breeder sales, with part of the stock going to their farm north of Craven, Saskatchewan. Bill, a knowledgeable cattleman with a long history of feeding and showing livestock, had an excellent record of preparing young farmers for their careers in agriculture.

Purebred shows in places like Regina, Saskatoon; Calgary, Alberta; and Brandon, Manitoba became regular events for us, along with Bettyanne's successful local 4-H club shows in 1984 and 1985. Although I grew up with livestock, there was little time then for 4-H clubs or showing cattle. I guess if my dad had not been required to come and help, I might have been able to join. Dad could not spare the time, even though Frank, a son of one of the first pioneers in the Longlaketon district, tried to get me to become a member and took me to a couple of meetings.

Close to our farm, the club's knowledgeable farm leaders discussed issues of grain and livestock production. Its noble mission was to instill good farming practices in the community's youth. Bill, my brother-in-law, had spent many years in the club and helped our family greatly with preparing purebred stock for show and sale. Each cow, bull, and calf required a name and to be registered by the Limousin Breeder's Association. Our hobby farm was expanding with the new cattle herd and grain operation. Our family adapted well to the exciting new reality of being a farm family.

—

Alberta's economic condition in 1984 was still not great. Canada Savings Bonds had a one-year, rate of return of 11% and a guaranteed 7% over a five year term.

In 1985, I was forty-three years old, residing in Balzac, Alberta, happily married, and blessed with three outstanding children. But I wondered about our future. It was going to be difficult to expand our farming in Balzac, land prices so close to Calgary and the high interest costs, *How are we doing? What's missing? What's needed?* One evening, Donna and I walked to the north field fence where Bettyanne had noticed a small break. It was better to attend to fence repairs when small and before the livestock broke out of our pasture.

"We're fine for assets, Doug," commented Donna. "With three successful businesses thus far, we have good equity."

"Yes, but cash is the problem," I replied, "Equity doesn't get the kids to summer camp or their 4-H Limousin shows. It is going to be hard to expand the farm so close to Calgary, I am confident we can succeed at farming, but we will have to grow and expand with the agriculture industry. I wonder, if with Dad getting older, and with his heart issues if it would make sense to move back to Saskatchewan, Darcy could certainly use help and we could work together and expand the family farm."

As is her custom, Donna said little. But I knew she was busy calculating and previewing the pieces required to make this shift: the schooling, housing, friendships, energy, and resources. I waited.

"It is humanly possible," she said, and I felt a twinge of excitement with the possibility of helping my Saskatchewan family and improving our family's future. I knew Donna, this city girl, would establish a great home for her family where ever we lived. Home is where you hang your hat!

We found the fence break, I repaired it that afternoon, and walked home. I was going to Saskatchewan in a few days to discuss joining the family farm operation. We wanted to build a new house on the home quarter and take our purebred cattle herd with us.

Dad was receptive to our idea, and we started planning the huge move. Despite the lack of financial guarantees, I was comfortable with Dad's fairness, and I think he knew I would work hard to help the farming operation. It was also important to us that Mom, Darcy, and Janet (Darcy's wife) were comfortable with our move. Yes, I had grown up on a farm, but then I left for university and began an entrepreneurial career. Sure, we had our Balzac farm with our growing cattle operation, but I knew this was modern grain farming and I had a lot to learn.

"Darcy," I said, "I'd be grateful for your advice. You have been a farmer since day one. You are way ahead of me in modern farming practices. Would you help me out?" With no hesitation, Darcy generously advised us about modern agrology, such as understanding weeds, insects, timing, and seeding rates. Dad helped, too, and was still quite involved. I think my machine servicing, repair, and driving abilities were a definite asset and contributed immediately once we got settled in the fall of 1985.

"So happy you're moving home, Doug!" said Mom, her happiness wonderful to hear. "My cookie sheets and storybooks are ready for my grandchildren."

MOVING BACK TO SASKATCHEWAN

DURING THE WINTER OF 1984-85, WITH A new Ford diesel three-quarter-ton truck and a blue trailer that Darcy built, we began our move to Earl Grey. We planned to have a new house built for the start of the new school year. In the meantime, Bettyanne, (Grade 8), Stephen (Grade 1) and Cathy (Kindergarten) would finish their school years in Airdrie.

I gave notice to Rocky View; I would be leaving the planning commission. I felt bad leaving at such a critical time in the county's development around Calgary. I had enjoyed the challenge, and it had been great to work with such a good group of people. I was going to miss them. I received a nice thank-you letter and was honoured by their reply: "Your standards of fair play were an example to all of us, and your incisive questioning and attention to detail was an inestimable aid."

It took multiple weekly trips to move our equipment and belongings. A custom cattle liner hauled thirty-eight cows, their spring calves, and one bull. The heavy equipment, including our new 2290 Case tractor, was custom hauled to the Earl Grey farm.

The trips went well, except for one in a terrible winter blizzard. It was time for calving, and as luck would have it, a cow was calving out in the middle of the pasture. Donna was desperate. With no cell phones, I pulled into service stations along the road to call her every half hour. At Drumheller, about an hour from Balzac, she said, "Hurry!" However, the RCMP had closed the highway to Calgary. A patrol officer suggested a backroad out of the valley. After plowing through drifts on country roads, with poor visibility, and following a snow plow for miles, I finally arrived home. Unfortunately, by the time I arrived, our pregnant Limousin cow had died from the stress and weather conditions. What a sad lesson in the ups and downs of farm reality.

Spring came, and I was helping on the farm and planning our new home. Donna was managing our affairs in Balzac. With school out in late June, we moved to our cottage in partnership with my sister and her family at Clearview, Saskatchewan while waiting for the new home construction to be finished. The cattle herd was custom hauled for the summer to the Robinson pasture along Last Mountain Lake, close to our cottage.

Our house was not quite finished by September 1, but we were excited to move, and school was starting, so even some missing floor covering, doorknobs, and moulding could not deter us. There were few dry years in the mid-1980s, and this year was no exception. After moving, we had to cut and bale roadside and sloughs for winter cattle feed. The grain crops were thin and short, with little to harvest. Cattle feed was expensive, and the method of double swathing into fifty- to sixty-foot swaths for combining and baling, thus salvaging more of the crop, was the most common. In September, the school bus picked up Bettyanne (Grade 9), Stephen (Grade 2), and Cathy (Grade 1) for the seven-mile ride to Earl Grey.

One day during that first harvest, I was riding with Dad, who was showing me how to operate his modern 7700 self-propelled John Deere combine. That machine was his pride and joy. "Doug," said Dad, "back in the day, before these air conditioned cabs, I could tell you how well my gas-run combine was running just by listening to it. Today, though, farmers cannot hear a darn thing way up high in these modern cabs. Bad stuff can happen well before an operator even knows about it"

"Sure, Dad, you used to be able to assess the sounds and what they meant mechanically, but there were other dangers. For example, you used gas, which is highly explosive, and your tractor's gas tank sat above the motor, close to the pulleys. I used to watch, Dad, how carefully you filled it, worried every time that excess might run down over the spark plugs and batteries and ignite. You told me that many farmers lost their tractors to fire that way and to never leave the engine running when filling with gas."

"Yes, I did. Today the fuel tanks on new machines are located way in the back, a good distance from spark plugs, and farmers use diesel fuel, which is not dumped in forty-five-gallon barrels onto a ground and hand pumped into the gas tank. As you know, prior to hand pumps, a pail was filled by tipping the barrel on its side and letting the fuel flow into the pail. It was then poured into the tank through a funnel. So, it is true Doug that with the large diesel fuel farm storage tanks, safety has improved."

"And remember, Dad, there were no fire extinguishers on our equipment back in the day. We lived with the threat that excellent dry harvesting conditions also meant more risk of fire." I remember pouring coffee out of my thermos onto a hot, smoking bearing. For that reason, I began each day with machine inspections and repairs before starting the combine. At the end of the harvest day, portable air blowers cleared the dry chaff and kept the machine super clean.

"I saw it with my own eyes," said Dad, "dry straw around bearings and shafts starting to smolder. That is the very reason I did a final check every night before bed. But the biggest grain farm risk of all, Doug, is the weather. It is destructive and merciless. Random farms in Saskatchewan still get walloped with years of back-to-back terrible crops. Picture the distress of one farmer who is hit by bad weather while, down the road, another is busy hauling in his bumper crop. That poor farmer could own half a million dollars' worth of equipment and yet not have the ready cash for the diesel fuel to bring in even the smallest quarter of his crop, all thanks to random lousy weather."

The jam-packed fall was taken up by harvest, building corrals, and moving the cattle back from the Robinson pasture. One example of the constant demands was how, one day, after rising early to prepare and show the animals at Agribition in Regina, Donna and I built our new corrals under the truck lights, long after the kids were in bed. The farm had virtually no fences, as Dad had sold his herd in 1970. Although Darcy had a few livestock before we came back, a great deal of wire fence was needed for our cattle. So, Darcy, Steve (our right-hand man from Calgary), and I installed five to six miles of new fence, which was a major undertaking. A cold winter was coming, and our livestock, home from the summer pasture, absolutely needed pasture and shelter.

Our first Christmas season was spent on the farm in the new house after a hugely disruptive year. On New Year's Eve, we all cleared a spot across the road at the dugout and had great fun skating. The snow was deep, but we soon built a fire and enjoyed hot chocolate and snacks. What a memorable winter's night for us all!

The next couple of years were taken up with farming, settling into Saskatchewan, the kids' school life, and stunningly, Bettyanne becoming a teenager. I worked on improving the farm and making a living while Donna was busy, again, finishing and decorating our new house, developing the yard, managing, and looking after the family. Even though we received some grain income from Dad for the first few years (1985-1988), our family's main income relied on the Limousin cattle herd.

The demand for our breeding stock with new developing herds increased our sale of Limousin bulls and cows. Darcy also raised cattle, so haying and fencing became a central part of the farm activity. The seeding outfit Dad had at the time was a thirty-three foot Morris Seed-Rite, pulled by his four-wheel drive 8440 John Deere tractor—a hoe drill and back rod to roll out newly germinated weeds; it was considered an early spring seeding machine, one of the few direct seeding tools of the day. Dad was one of the first few farmers to use this type of seeder due to farm help shortage in the spring and his desire for early seeding. The problem was that, while it worked in preparing cultivated land, it did not work on stubble land that was cropped the prior year. I had seen stubble cropping in Alberta, and with proper fertilizer, I strongly believed that continuous cropping was the way of the future.

Dad's tradition, and that of many other farmers, was to let the land lie fallow every second year. He believed this controlled weeds and replenished moisture for the following year's crop. Thus, after forty years, Dad had harvested only twenty crops. This was the common practice. This habit of two-year rotation had served him well through the post-war years of the 1950s to the 1980s. The mid-1980s were dry years, and instead of revitalizing, the soil was blown away by the winds. The weeds absorbed all the moisture, leaving the ground dry and uneven, and when it rained, there were water ditches and soil was washed away.

To me, the cherished notion of leaving the land lie fallow made no sense and was outdated. When the choice was mine, I began building equipment for direct seeding into stubble. In this way, the seeds had moist soil in which to germinate and keep growing, conserving our limited moisture. I also helped improve our spraying methods for controlling weeds.

A couple of years after we arrived, Dad, Darcy, and I purchased the Hatton farm (640 acres) with four quarters on Highway #20, west of our home farm. Mom and Dad soon moved to the new Hatton farmyard for the summers, so they would be on pavement and not muddy roads. They also had a lovely home in Lumsden, where Mom loved her sporting activities. It seemed that Dad did not enjoy the cattle herd, or my new farming methods, because of the amount of change. I think that was the beginning of my strained relationship with Dad.

Along with maintaining the current farm equipment, I started manufacturing new and useful farm equipment with the help from our neighbour, Al, and welding friend, Marlin. For example, I built a post pounder on a trailer; it would pound posts into the ground for building cattle fences. We also fashioned a ten-foot landscaper from used parts. The scraper was never painted, which I still regret every time I see it, but is used to this day.

The seeding issue continued to be a priority for me. I purchased a used John Deere 65 air seeder and proceeded to convert it into a no-till drill by adding multiple new seed boots for seed, fertilizer, and on-row packers. The grain tank was on the hitch, which was a disadvantage when full and in wet conditions. However, it allowed the anhydrous tank, filled with liquid fertilizer, on a four-wheel trailer to be towed behind the seeding outfit. This seeding process change took place over the next four to five years, starting with granular, liquid, and gas fertilizer. I enjoyed studying continuous cropping and the effectiveness of fertilizing and other inputs on the crop.

I am glad I was part of this farming history. I think the idea of no-tilling heavier producing crops and saving the fertile ground layer has helped dramatically to save our environment. The control of weeds, with the improvement of chemicals in the late 1980s, meant spraying equipment had to be improved and made much larger.

In 1952, Dad purchased his first sprayer, which bolted to the front of his John Deere B tractor. The tank sat on the back hitch, and the spray sections totaled thirty feet and cost approximately $400.00. The only chemical used was 2-4-D for weed control; the original formula is outlawed today.

I built a one-hundred-foot pull-type sprayer out of a forty-eight-foot manufactured rod weeder. I purchased it from Hodgins Auctions in Melfort, Saskatchewan, and Stephen and

I pulled it home, which took much longer than I planned due to the size and flopping castor wheels that limited our speed. At age nine, Stephen was not impressed with my purchase, and he still talks about the long and slow trip. Our manufactured sprayer, folded for transport, had chemical and mixing tanks and a second boom for double spraying of two types of weeds. Soon, however, with the fast-moving times, Douglas Industries Ltd. purchased our first self-propelled sprayer for our farm and did custom work for other farmers. My home-made pull-type sprayer was no longer needed.

The Earl Grey area had more than its fair share of field rocks, which caused equipment damage. There was much hard labour, especially before rock-picking equipment was invented. As time went by, the horses were replaced by a John Deere B tractor, which was a marvel to me and a great little machine. The larger hand-picked rocks were rolled onto the wagon and were later thrown off onto a pile beside a slough or, if too big, onto a chain that was attached and drug off into a bush with a team of horses, or a tractor. The first rock picker, called an Anderson Rock Picker, was designed in the next town, Southey, and came into use around 1960.

On returning in 1985, I built a rock fork, with wheels to pull behind a tractor to remove larger unmovable rocks that had been left in the ground causing equipment damage, sometimes called dead heads. However, finding them was difficult, so I began designing and testing various methods of marking these partially hidden rocks when we were seeding or harvesting and running over them. I applied to Prairie Implement Machinery Institute (PAMI) for a grant to help cover part of the expenses in the design and development. They researched, improved upon, and tested my initial prototype. The result was an automatic flagger (a two-foot wire with a red flag on top) dispensed by a machine attached to a swather or cultivator. When operators saw a rock, they would flip a switch. However, once again my timing was bad, because even though we were close to a useable product, technology made our mechanical device obsolete.

On arriving at the farm, the harvest equipment comprised of two single three-ton axle grain trucks, Dad's self-propelled 7700 John Deere combine, and Darcy's 6600 John Deere pull-type combine. With Mom's supervision, Janet and Donna drove the grain trucks. Dad managed the unloading and prepared the grain bins for the first few years because of all the intricate of filling his bins, especially the wood bins, which only he understood.

Donna so marvelled at his number system for the six-inch door boards, the exact auger placement, and the proper back-up angle, along with how final shoveling was needed to properly fill and maximize the amount stored. He always had the bins ready and was there in case of any problems the truckers had unloading. Bettyanne was in high school now, so she helped with the cattle, while the youngsters (Stephen and Cathy, and Darcy's children, Chelsea and Pam) rode in the grain trucks. These were good family times, especially when we all stopped to eat supper. Mom's hearty stews, Donna's delicious pies and ice cream (Dad's favourite), and Marleen's great cooking kept us all well fed. I always joked with Marleen, a particularly good friend of the family, about getting a chuck wagon to follow us around at mealtime.

—

In the late 1990s I had another idea to reduce the harvest period time. Moisture had returned to our area, and with Stephen now able to drive our tractor and combine (with Darcy's supervision from his outfit), I purchased an older self-propelled John Deere 6600 combine with a twenty-four-foot straight cut header. From a local farmer, I obtained an older cab-over semi. We fixed up and installed a roll-up tarp cover to protect the grain in the trailer when it rained. Thus, I was now able to combine by myself for quite some time with the twelve hundred-bushel truck storage. In straight combining, the advantage is you do not have to follow a swath around the field but can work and cut through a crop for convenience and unloading.

—

A couple years after our move, I was ready to volunteer my financial expertise with the agriculture community in some small way. I wrote to Premier Grant Devine, of the province of Saskatchewan, who was a first-class agricultural economist I had met when he was a professor at the University of Saskatchewan in the 1960s. I asked him if I might volunteer to be of service to the province.

The letter I received back was from Farmland Security asking me for an interview. Soon I was appointed to the Farmland Security Provincial Board. I attended monthly board meetings and creditor/farmer financial negotiations that addressed the farm crisis taking place. I enjoyed analyzing financial figures, helping the board follow government policy, and giving farmers financial help. John, a personable man, helped me to develop my skills in negotiating settlements. He became a good friend of mine through those times. I had great respect for the group and especially for Ken and John for their leadership and understanding of my direct control-management techniques that I had practiced for most of my life. They appreciated my trying to adapt to the new group decision-making process.

I met a great bunch of professional part-timers like me. The government continued to require more of my time, and the income meant a few more family holidays. There was a great deal of province-wide travelling and long hours of juggling farming and personal schedules. On one occasion, I sprayed a tank of weed chemical before 6:00 a.m. and had a meeting in Weyburn at 9:00 am.

—

"Mom! Dad!" shouted Cathy, her voice filled with dread. "The barn roof is flapping off! It's lifting in the wind!" I was wearing my blue business suit, and by habit, grabbed my car keys in my rush to the front porch. In three hours, I was supposed to be at my meeting with the Farmland Security Board, a quasi-judicial tribunal whose members are appointed by the Lieutenant Governor's Council, as I was. I glanced to my left and saw the barn roof was indeed hanging on for dear life. The six-inch screw shank nails were proving their worth in this category 4 Saskatchewan windstorm. One hundred and fifty kilometre per hour winds committed a last violent assault as the roof lifted higher. I note, with gratitude, that our newly restored fence was mercifully spared.

The roof settled back down on its mounts as the gusts subsided. "I've got this, Doug," said Donna. "It's not a fire."

My children looked at me with eyebrows sky high and then looked at their mother, whose worst fear for the farm was fire.

Farm equipment generates a great deal of heat, which can easily spark a fire, but this flapping roof was not in flames.

"You better head out," said Donna.

"The animals!" Cathy interrupted, tearing out to the barn with Bettyanne in hot pursuit and me yearning to follow my daughters.

"Dad?" asked Stephen.

I felt split. One leg stepped toward the barn. The other pointed to the car. The roof was in danger of blowing off. In three hours, though, four farmers expected to meet with our council, and great efforts had been made by all to attend this meeting.

"I have an appointment, son," I said, as I had said many times before.

I could not be absent. I argued with myself. Scattered thoughts flew like ripped waterproof, asphalt-coated shingles. I could not toss four farmers into the air like a sheared roof. No—I said I would be there, so I'd be there.

Late the night before, at the kitchen table, I had reviewed each entrusted file once more. After a month of reflecting about each situation, I had a surprise strategy to share with one couple: a small but helpful loophole to rescue their mortgage. Another farmer was desperately concerned about the foreclosure protection he believed he was entitled to, while another couple was distressed about an unprecedented technicality in their home quarter protection. Two investors had flown in from China to discuss critical details about regulating the ownership of their Saskatchewan farmland with the Farmland Security Board. Two weeks earlier, I'd revised one of my suggestions after discussing the matter with Donna.

My wife and I were never in contest; we did not compete with one another. I sought out, and was grateful for, Donna's opinions.

Donna's ideas and original perspectives, sometimes so different from mine, are most valuable to me. I love the wild card she often produces—the unsuspected consideration, the unique angle—and as such, Donna enriched many of my ideas. I still appreciate that, together, we have assisted many farmers, whose work is never easy.

So, there I was on the porch in my suit, roof waving, kids upset, and animals anxious, fearful of safety and cost issues.

"Go," said Donna, pointing to the car.

I pulled open the driver's door—careful of the wind ripping it from its hinges—sunk into the seat, tucked my water bottle into its cradle, and drove out of the yard. The wheels crunched the gravel, making a sound that stirred a toxic combination of guilt and helplessness in me. I was torn about my choice yet compelled to show up on time for the strangers who could benefit from my business experience. Those folks had families to feed just as I did.

I glanced at my watch; another hour to go. Stephen, Bettyanne, and Cathy had undoubtedly fed the animals. Roof or no roof, I knew my kids had checked-in on our two pregnant Limousin cattle and had them safely positioned in a squeeze holding pen in preparation for calving. Donna had checked that the roof was finally secure since our hired hand's truck drove in just as I left.

I took a deep breath. My family had it under control. I glanced at my sun-warmed briefcase and took another swallow of water—liquid gold unavailable in Saskatchewan during the Dust Bowl days. I arrived in the parking lot ahead of schedule. There was a wide-open space for me.

> PREMIER OF SASKATCHEWAN
> LEGISLATIVE BUILDING
> REGINA, CANADA S4S 0B3 (306) 787-6271
>
> On behalf of the Government of Saskatchewan, I join in honouring Doug Brewster for the devotion and attention given to the accomplishments of the Farm Land Security Board.
>
> In this period of our history, Saskatchewan has witnessed moments of sadness and difficulty in the farming sector. In administering the responsibilities of the Farm Land Security Board, Doug Brewster has displayed wisdom and kindness, and an abundance of character, tolerance and worth. The people with whom Doug Brewster has worked have all benefited from the dedication to excellence shown by Doug.
>
> We join in wishing Doug all the best for the future. We wish Doug rich and deep happiness, good health, prosperity and friendship.
>
> Our sincere thanks for all that you have done for Saskatchewan.
>
> Grant Devine
> Premier

A year or so after my mandate on the board, the Department of Justice set up a new department called Mediation Services, which was intended to mediate court actions disputes, such as farm debt and family break ups. This was like what we had been doing but with mediation techniques of neutrality, and the added areas of dispute were entirely new. Ken, head of Mediation Services, asked me to join and get educated as a mediator to represent Mediation Services, so that I could mediate dispute matters assigned by the courts. I continued to work from home and travel to meetings, with Donna's help in handling my correspondence with clients, lawyers, and creditors. The introduction of our first Apple computer and Donna's administration skills made it all possible.

During 1988 to 1990 period of working for the government, I received one of the first portable phones. Now I could keep in touch with Donna and the farm. The phone comprised of a handset (like a house phone), a battery, and a charging unit, all installed in the car trunk. It helped greatly, even though it was bulky and crude. How far we have come since then!

One day, arriving home from a meeting in Saskatoon at 9.30 p.m., I saw Donna out front waving to me, but this wasn't a welcome-home wave. It was an urgent alert she was signaling while she dashed to the barn. Three cows were calving, and I sensed a problem with one. Within minutes I was on the phone to Debbie, our trusted vet, who has saved the day for many of us in the Earl Grey district. We soon welcomed the first healthy newborn Limousin calf, but an hour later we grieved the second newborn, who laid lifeless before us, despite our every effort. The final calf was born healthy and already trying to stand. (It is important that newborn calves are quickly dried off, standing, and receiving their mother's first rich milk within a few hours of being born.) By 1:00 a.m., the day was done and a new one had already begun.

I had noticed a braided rope hanging in the barn with strands of purple, green, and blue. It caused me to reflect that our days were woven with shock, service, ruin, birth, and death. In our grief, we all knew we had cooperated to weave the unpredictable into another day well-lived.

"I left the files you need to review for tomorrow's meeting, Doug. They're on the kitchen table," said Donna. Then, the two of us fell asleep.

—

Our kids were educated in the Earl Grey community school, and Donna and I became quite involved. I became a member of the local school board. Bettyanne graduated in 1989 and headed off to Saskatoon to earn her veterinary assistant diploma from Saskatchewan Institute of Applied Science and Technology (SIAST) training at the University of Saskatchewan Veterinary Collage. Both Stephen and Cathy were still in school in the Earl Grey when we heard rumours of it being closed. We did not want our children transported to the Southey school, and many other parents felt the same. That would mean an extra hour of sitting on the bus each day. A local group of supporters was formed, and we worked together to ensure our school remain in Earl Grey. We contributed funds to obtain legal advice and promote the Earl Grey School. After all, a school and its students are an integral part of a community. They bring life and vibrancy to their community.

With the help of the group, particularly Laurie, our organizer, I agreed to run for the Cupar School Division Board, in favour of keeping our community school. I succeeded, and due to several changing circumstances and other new board members, the school closures in Earl Grey and Dysart were reversed. We did it! It took hard work, reality checks, and luck.

The administrator at that time became ill and was bed ridden, so the assistant, Sandra, took the job. I was asked to review the past accounting records of the division, where I discovered a number of over-spending problems. We found many ways to cut operating costs for the division without affecting Sandra's drive for excellent community education. We worked well with the board. Our group improved the quality of communication and education in the division, along with cancelling the proposed education tax increases.

Unfortunately for us, Sandra moved on a couple of years later, and the new administrator made many drastic changes. In my opinion, his position on many issues was not in the best interest of our communities or our children's education. Jack, another board member, and I were upset with the issue and decided to leave the board. In retrospect, we should have stayed to fight another day, but I felt we had lost the battle. I received a lovely letter from Sandra, now the director of education, upon leaving the Cupar Division. It read in part:

I will take many fond memories with me as I leave the Cupar School Division. I appreciated your intellect, your sound financial sense, and the ability you provided me and the Board of Education. You have been a significant part of the leadership that has brought about positive changes and improvements for the students, parents, ratepayers, and staff in the Earl Grey area, as well as the Cupar School Division.

A couple of years later, the Earl Grey School closed, and students were transferred to the Southey school. This action was detrimental to the future of our community, as well as our local sports and business.

—

Like Bettyanne, Stephen and Cathy became involved with the livestock. All three of our children did well showing and grooming and were a huge help in feeding and training our Limousin cattle. Donna and I could not have operated our cattle operation without their valuable assistance. They were committed to the 4-H Club, showing Limousin cattle and winning their proud share of ribbons and trophies, thus helping our bull and cow sales. The height of Bettyanne's impressive accomplishments was to represent the province of Saskatchewan as Limousin Queen.

Stephen joined Sea Cadets, where he thoroughly enjoyed learning and participating. Donna drove him to and from Regina every week. In Grade 11, our son was chosen to go on a three-month work cruise aboard the HMCS Regina, leaving from the Canadian Naval Base CFB Esquimalt, located in Victoria. The trip included visiting Pearl Harbor, playing war games with the United States Navy, and visiting Hong Kong and Australia. He had the great opportunity to cross both the international dateline and the equator.

Cathy played piano and loved singing, both talents we admired. Indoors or outdoors, our daughter wasted no time; she was eager to attend summer camps in Prince Albert and Humboldt. In a six-week Big River summer production, The Adventures of Huckleberry Finn, Cathy helped create the original set and sang in the choir.

All our children were involved in various sports, such as curling, volleyball, rugby, football, and wrestling, which we supported. Unfortunately, Donna usually wound-up with job of driving them to local communities, Regina and farther. I really do not know how she did everything.

—

I began to feel the pressure and strain of travel in the early 1990s. It also became clear to me that family mediation was not of keen interest or my area of expertise. Even with Donna's help in driving and typing reports, the workload and stress forced me to slow down. I felt bad about leaving Ken, John, and the mediation organization that had been so good to me, but it was time.

—

As Stephen and Cathy became older and more self-sufficient, Donna spent a couple of winters studying income tax with H&R Block in Regina, very much enjoying the challenge. Always involved in our accounting and tax preparation, Donna was top notch at accuracy and punctuality. My procrastination over the years drove her crazy. I have claimed professional accounting students get dipped into a tub of some sort of solution that causes stalling to the last possible second on absolutely everything, especially our income taxes.

Soon, Donna worked full time every spring during tax season and thoroughly enjoyed her team at H&R Block in Regina. For a great many of us, Donna was the go-to girl for current date tax law. Donna prepared returns for friends and family every spring before 12:00 p.m. on April 30. To this day, I dread early May's cold shoulder. I always ask Donna to do our tax return, after getting her all the information necessary right before the deadline. Frankly, delivering our tax returns at 11:30 p.m. on April 30 was common for us when we lived in both Calgary and Earl Grey. On a few occasions, Donna would meet me at the Earl Grey post office to get my signature.

—

A new method of preparing bale feed for livestock appeared in the late 1980s/early 1990s was with a machine called a shredder. It could self-load and shred large round bales, providing better nutrition value for animal consumption. It had a set of wheels and hitch that were pulled and driven by a tractor's power take off, dispensing a swath of chopped straw alongside as it moved forward.

At the beginning of its development, the machines on the market were expensive and, it seemed to me, could be made much simpler, affordable, and available for smaller stock operations. Harold, a friend of mine, had a small manufacturing facility in Strasbourg. Our history of working together in manufacturing and selling product went back almost thirty years, so I approached him for help in manufacturing my new design. Wayne, an excellent welder and fitter, worked for Harold. Before long we built a prototype of a new, simpler, and less costly style of shredder. The initial unit was crude and not visually appealing, but it performed well in the testing stage.

However, Harold's operation in Strasbourg was moving to Grenfell, and I had to decide whether to proceed with the manufacturing or shelve the idea. In the end, we decided to move to Grenfell with Harold's shop, where we started manufacturing the Douglas Industries Ltd. 202 Shredder. Douglas Industries Ltd., our Alberta-based construction company, was established in the 1970s for general contracting and developing. It became active again in Saskatchewan as an equipment manufacturer.

202 SHREDDER

Harold began selling the 202 Shredder, and I worked on the design, drawings, and material purchases. I travelled the three-hour drive to the plant a couple of days a week to help in the production and painting. Despite a great deal of demand for the product, we had our many challenges of growing pains, such as perfecting the product, marketing, warranty issues, and needing more capital.

In the meantime, Bettyanne completed her veterinary program and found a job working with livestock at Pound Maker, a large feed lot and ethanol plant near Lanigan, Saskatchewan. She married and eventually moved to Yorkton, Saskatchewan.

Stephen graduated Grade 12 and was having fun at Lakeland College in Vermillion, Alberta. Cathy had also graduated and was at university in Regina, taking business administration and surprising Donna and I by joining the Army Reserve without discussing the matter with us.

—

We decided to sell the lake cottage at Clearview, which my sister and her family half-owned with us. The summer cottage had been a great place for our growing young families, and we had many wonderful times and memories there.

I purchased two quarter sections of pasture beside Last Mountain Lake, west of Strasbourg. Our plan was two-fold: to create a summer pasture for our livestock herd, and to create a summer cottage on the lakeshore. Initially, we used a little travel trailer and soon moved the small insurance building from Earl Grey. Donna planted lots of trees around our new cottage. John, a friend of the family, rototilled the yards and helped a lot in controlling the weeds. The electrical power was brought to the site and water was pumped from the lake, making our little place quite enjoyable. Our family spent many hours enjoying and working at the property. I found this isolation rejuvenating and wish I had spent more time absorbing these precious moments.

We had our share of fearless thunderstorms coming off the lake from the west, especially night storms. We would lay in bed and listen to the wind, waves, and thunder, and we would watch the lightening glow across the lake. Only we and our small fortress sat on that vacant beach, defying nature. Those were moments one could never duplicate.

Donna and the kids all enjoyed the lake and the beach, as did I, and this remote location with its natural beauty only added to our pleasure. We built the pasture fence and loading corrals for our Limousin herd. The location was great for the cattle in the summer months, with lots of fresh water, lush grass, and the unusual fact that we could check on them as we relaxed in our boat on the lake.

Once Stephen and Cathy left for college and university, we sadly sold the cattle herd and our summer cottage and pasture. A good part of the herd was sold to Darcy, and the rest was sold to a fellow from Strasbourg. Although we missed both the cottage and the Limousin cattle, Donna and I knew we had to reduce our workload and responsibilities.

The grain farm areas continued to increase, with excellent help from fellows in the area. John, after retiring from Saskatchewan Highways, spent many years helping us from spring seeding through harvest. I enjoyed improving the land, and John helped greatly in filling potholes and leveling hilltops.

Farm equipment had improved, making life much easier for our family. I bought larger equipment and continued my involvement in designing and manufacturing farm equipment with Douglas Industries Ltd. In 1995, I rented and later purchased a property in Earl Grey (107 Railway Avenue). It was a service station built in 1961 that I felt could be adapted for manufacturing the shredders. Lyle, a local farmer and neighbour, said he'd help me get the old shop cleaned out and converted over for manufacturing. I purchased equipment and hired people, and we were into production in short order. It would not be long before we needed to expand and adapt the shop to our needs. I would have never guessed this was to be the start of a facility that would serve our family's needs for many years to come.

Our 202 Shredder had proven to be a big success. However, another Saskatchewan company, High Line Manufacturing, felt our product design infringed on their patented shredder design. After investigation and legal advice, I decided not to contest their claims and settled out of court.

We began work on a new design for a sliding table over a rotating flail, rather than bale turning. With Wayne's skilled help, we built a prototype in his Strasburg shop. After, some months of field-testing, along with implementing changes and improvements,

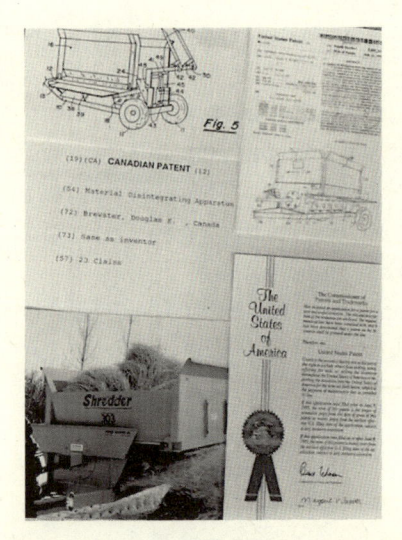

303 SHEDDER

production began. I asked Gil, a retired engineer who grew up in Earl Grey, to help us work through the patent process. Gil lived in Regina and helped inventors with their patent applications. It took a great deal of time, effort, and resources, but in 1995 we obtained both the Canadian and American patent for our design.

Our Earl Grey manufacturing facility was a busy enterprise when I designed the new 303 Shredder and a self-unloading bale wagon. Sales were strong. However, with some shredder design issues and insufficient financial resources to remain competitive, we fell behind our competitors. Our location was ideal for local sales and service for the surrounding communities. We had been selling our manufactured bale wagons, flax bunchers, and shredders, as well as many other things. We started renting agricultural equipment, which soon became a big part of our business. Designing and renting equipment, such as land rollers, grain carts, and flax bunchers, helped us make our living and aided us in perfecting our machine.

We had superb staff in Earl Grey's manufacturing shop, although I was always busy with our farm. Carol, who I'd known from my high school years, was a great help to me. In 1993, I called her and asked for her help in the office and shop. Carol drove the school bus and had five hours between her daily bus trips. She and Birney and their family have a mixed farm north of Earl Grey. I knew her work was never done but asked anyway. Carol spent the next six years with us and was a huge contributor to our success. From helping Donna with the accounting, putting labels on new machines, cutting steel, and being girl Friday, Carol did it all. In Earl Grey's 1995 parade, Carol did us proud by creating our Douglas Industries Ltd. float on farm progress, and I showed off my idea of a self-driven grain auger. One year, Donna set up a surprise birthday barbeque lunch for Carol's birthday. Carol was appreciative and had a great time with all the staff.

WANTED

"JACOB SNEAKY HERMANSON"
FOR QUESTIONING IN RECENT
BREAK-INS, DISTRIBUTION OF
PETITIONS AND FORGERY.

ANYONE SEEING THIS
SHADY INDIVIDUAL: PLEASE
CONTACT LOCAL AUTHORITIES.

HE IS KNOWN TO KEEP
COMPANY WITH
"BIG GUY BRUCE HORNBUSH"
ALSO WANTED FOR
QUESTIONING.

ANYONE SEEING THESE DUBIOUS
DISREPUTABLE CHARACTERS, CONTACT
AUTHORITIES IMMEDIATELY.
DO NOT APPROACH THEM AS
THEY MAY BE DANGEROUS.

We had our fun times. Carol was the instigator in most cases, and she could certainly hold her own with the guys in the shop. She had many talents and was one of those quiet, unsung people who may never get their fifteen minutes of fame. We had some outstanding people in both the office and shop, including family members. There was Karen, Donna W., John, Craig, Danny, and Garth, to name a few, along with many others who helped keep things running smoothly.

Drawings and layouts for cutting and assembly were important. Perry, a local friend with training and experience, helped greatly with these areas of the manufacturing. Perry and Donny, another local fellow, helped me from time to time with the viability of new ideas. I so appreciated having input and factual criticism from all involved.

The equipment rental business was booming in the mid and late 1990s, because farmers could not afford $30,000.00 for a machine they used for only a couple of weeks a year. The area around us was still a mixed farm area of grain and livestock. There were many part-time operations between us and the fifty kilometres to Regina.

—

As time passed, we became grandparents with Bettyanne's girls, Jessica and Callie, and their brother, Luke—the cutest kids in mine and Donna's world. Stephen studied in between having fun at his two part-time jobs in Vermillion: one as a lifeguard and one as house-dad at the co-ed dorm where he lived. He later got a job with the Royal Bank in Moosomin, Saskatchewan. Our youngest, Cathy, was off training in the military while earning her administration degree at the University of Regina.

I felt strongly about grinding weed seeds for livestock feed as an extension of our operations. I researched facilities that not only prepared the grain dockage weed seeds, cracked grain, and unthrashed heads but also pelletized the final product. The dockage, a by-product of our grain product (worthless at the time), cost the farmer handling fees on shipment. The idea really took root when it became obvious that the old local grain elevators were going to be demolished, thus providing a used grain cleaner and necessary piping materials. Also, an electric grain grinder became available at a local farm auction; Carol's husband, Birney purchased it for us.

Consequently, a second location for grain cleaning and grinding, grain storage, and liquid fertilizer tank storage was established in Earl Grey. Although I intended this new facility be used for our farm, it also added revenue of feed and liquid fertilizer sales. I purchased four acres of bare land next to the old wood grain elevator site from the village of Earl Grey. It is now in the history books and near to our Earl Grey shop. The last two grain elevators were closed and dismantled, with a new large grain terminal built between Earl Grey and Southey. The welded structural steel intended for the cleaning plant was processed at our shop. It would accommodate a seventy-foot vertical grain conveyer and two grain cleaners: a large used cleaner from the old Earl Grey Pioneer elevator that was demolished and one from Flaman Sales in Southey.

Today the business, Flaman Sales, is spread across western Canada and continues to supply and service the agriculture community, as well as the fitness community. I marvelled at Frank Flaman's ability to figure out what a consumer needed and to provide it for them at an affordable price. To me, it seems all about the price he paid and a reasonable mark-up for success, which I always tried to practice.

Unfortunately, although the new facility had become an intricate and important part of Stephen's farming operation, my initial idea never came into fruition, due to my pending health problems and our farm expanding beyond my wildest dreams. This facility and yard has been a huge undertaking, with many hours of leveling ground and constructing the cleaning plant with a grain leg, bin storage, etc. It included the involvement of our employees, as well as local people and businesses, an example of which is the time when Rudy, a local businessman with a great deal of building metal, experience, and a crane truck, got involved. His knowledge and contribution to the area, our projects, and others throughout the years is unmeasurable.

LIFE'S REALITY

I BEGAN TO DOWNSIZE OUR EARL GREY manufacturing operation in the late 1990s. I wanted more leisure time to spend on the grain farming operation. "But fate had something different planned for me and my family.

Dad was now well out of worrying about the day-to-day goings on. He liked his quiet tradition of meticulous maintenance, along with picking rocks using his Degelman rock picker and 4240 John Deere air-conditioned cab, residing in his Hatton farmyard on Highway #20 in the summer. In winter, he and Mom lived in an apartment they had near the North Gate Mall in Regina. He would refer to me as a businessman and not a farmer, and I sensed resignation on his part. Dad, well into retirement, was not interested in change of any kind; he was content with his accomplishments, of which there were many. With me, it was all about production and moving ahead, but Dad was happy with any repairs to the existing yard and equipment and would have been much happier if I'd have stopped there. Still, in looking back, I'm glad I was not dissuaded from being who I am, as we'd have not survived the changing times presented to us in the last fifteen years of the twentieth century.

As the fall of 1999 ended and the millennium approached, everyone anticipated many changes. I never imagined our family life would be jolted with such a reality check at the start of the year 2000. Late that fall, we received a phone call. Dad was in the hospital with a heart problem and needed a pacemaker. On arriving at the hospital, we were told that after he returned from his pacemaker surgery, he experienced a severe stroke that paralyzed him. The family was devastated and had to begin dealing with his medical and business issues. It was so difficult to see our dad in this condition—he was unable to move, and we were unable to make sense of his speech.

A few weeks later, we learned that my sister Donna's oldest son, Michael's life abruptly ended. It was a terrible shock, as he had been up visiting Dad at the hospital only a few nights prior. I can still see him pointing out the hospital window to show us his beautiful new black GMC truck. Michael's sudden death was a blinding blow to everyone in our family.

Our much-anticipated millennium Christmas was too quiet instead. We met at Cathy's house in Regina for turkey while Dad remained in the hospital needing total care.

We worked as a family to adjust to our new reality and deal with past and present matters. It was challenging, but perseverance is the power to continue. In June 2000, with the seeding done, we received a call that Dad had passed away in Wascana Centre. His nurse said that with his immobile condition and communication difficulties, he just gave up. Mom had been living on her own and doing well—she was quite active—but losing Dad took a huge toll on her, and Mom passed away five years later, in 2006.

—

The roots of the willow tree are remarkable for their toughness, size, medicinal quality, and tenacity to live. I find it fitting that in the year 2000, they formed the background for setting my life's new course.

In the fall of 2000, Bettyanne and I planned a fun weekend in Strasbourg, Saskatchewan to learn the fine art of willow tree furniture-making. I drove over a day earlier to cut and stack the branches needed for the course. Despite the unusual fatigue I felt while working that morning and ignoring the fact that I needed to lean against my car near the slough for a few minutes, I carried on. Driving home that afternoon, I felt unusually warm. I lowered the window for some cool, fresh air. At home, I took a couple of Tums for indigestion and laid down on the bed. Unable to relax, I wound up on our living room couch.

When Donna got home, she took one look at me and called an ambulance while I chewed two blood-thinning aspirins. After what seemed like forever, the ambulance arrived and connected me to much needed oxygen. Within minutes, we were on our way to the Regina General Hospital, where Cathy met us. I was diagnosed with having experienced a major heart attack. I'd need an angiogram immediately. The doctor explained the risks, and Donna asked for five minutes for us to decide.

"He doesn't have five minutes," was the doc's reply.

In mere seconds, gurney wheels raced down the long hallway with my doctor and his staff close behind. An angiogram showed there was a severe blockage of blood flow, and I needed a stent to open the artery, which they inserted immediately.

In recovery CCU, I spent the next long hours lying flat on my back to ensure the incision into the artery in my groin would not open and bleed. That excruciating procedure no longer exists.

Today, with thanks to medical researchers and advancement, an artery in the wrist can be used, and twenty minutes of a nurse's hand pressure in recovery does the trick.

I was extremely sick, and we soon learned the seriousness of the situation. It was life changing. Donna's sisters came to help support her and were in my room at the meeting with Dr. Zimmerman, who standing at the end of my hospital bed, told us he would not have believed anyone could have survived this heart attack. I was in for a long recovery and was told that none of the damage would ever repair itself.

With the recent deaths of Dad and Michael, it was a terrifying time. The future seemed bleak.

In the next couple of weeks before leaving the hospital, I was examined by specialists and began medication that had many side effects. Once home, progress continued to be slow as we dealt with business and family issues over the next months. I continued to recover and adjust to my prognosis of heart disease. Excellent doctors examined me regularly to adjust my medications as needed. My permanently damaged heart continued to pump, though I needed rest. Other than some legal and accounting work at home, I did little. On one occasion, Dad's estate lawyer in Regina drove out on a snowy day to have me sign documents. He got stuck in a snowbank on our driveway, and I could not even help him get his car back on the road.

—

February brought good news. Bettyanne's third child, a boy she named Luke, had arrived! Stephen moved to the Royal Bank in Strasbourg, so he could be closer to home and help on the farm. Also, Brian, a retired Saskatchewan wheat pool elevator agent, and our friend John helped take care of the farm work for us. Donna and I were so grateful to our friends and family for their help.

The next couple of years, Donna and I adjusted to part-time retirement. We travelled with our little motor home in the summer, and aside from short trips, spent the winters in Airdrie, Alberta in a rented apartment. Bettyanne was busy being a mom, Stephen had now taken over running the farm, and Cathy was travelling with the Army Reserve while completing her studies at the University of Saskatchewan.

My sister Donna, living in Regina was still grieving the loss of her son Michael and her marriage ending. She continued with the day to day, spending extended periods of time with us and being in and out of the hospital.

Slowly, I got back to helping on the farm a little, supervising the completion of the seed cleaning facility, driving equipment during busy times, and helping with the rental and sales of Douglas Industries Ltd. I was able to clear and improve our yard and agriculture land, playing with my big-boy toys as Donna likes to call them. I was unable to do the physical work myself, but it was amazing what a John Deere skid steer could do to clean out and remove the old buildings and unused cattle facilities, and expanding the yard for the larger farm equipment. We purchased a D6D cat dozer and land scraper for me to clear brush and level the yard and farmland. I worked on my own, so I could start and stop depending on how I felt. I enjoyed it a lot and felt lucky to be back doing something I enjoyed.

Donna did not return to H&R Block after 2000 out of concern for me, which I felt terrible about. However, she continued to be the one to call for tax questions, and she still prepared family tax returns. Donna also enjoyed her large farmyard and worked hard at landscaping. She would spend hours out in the hot sun riding the power lawnmower. In the past, mostly in spring times I would get in trouble for scraping her grass and leaving rocks around after I pushed snow with the tractor and snow blade.

—

The year 2001 was upon us. Everyone was busy, and the grandkids were growing like weeds. Even Luke, with his big smile, was walking trying to keep up with his two sisters. Summertime was when they would come and visit Donna and I, or Nana and Poppa, as we were known to them. They would stay overnight, and it was always great. Bettyanne and her family lived in Yorkton, Saskatchewan, a two-hour drive northeast of Earl Grey, so getting together took planning and often involved an overnight stay.

During my recovery, I decided to see if the main road north of us, where Fosterdale School had stood, could be named Fosterdale in honour of the school's history in the area. I wrote a request letter to the Council of the Rural Municipality of Longlaketon #219. I offered to fund the signage at the corners of Highway #20 and Grid 641, if they would allow, install, and maintain them. Our rural municipality at the time had few named road signs, but I thought this one would provide direction for drivers who were unfamiliar with the area.

The proposal was accepted, and within short order we had the permanent signs installed. They continue to honour the school and point the way for strangers to this day.

In June 2004 Cathy completed her administration Degree at the University of Regina and continued with the Canadian Army Reserve. That fall, Stephen and I found ourselves leaving work and rushing to the Regina General Hospital, where we met Donna as she was going into surgery. I appreciated that Donna's sister Susan, and Cathy, would not let Donna go in until we arrived, as only the two of them could have arranged.

That autumn, along with the falling leaves, Donna's energy had dropped. The decrease in energy scared Donna enough to visit our family doctor. The next day, a stress test revealed a heart problem, so she was sent to emergency. An angiogram showed a ninety-nine percent block, so a stent was quickly inserted, which allowed the blood to flow.

Once again, we were incredibly grateful for the fast action of the medical system and staff. There was no damage to

Donna's heart, and after a few weeks of rest, life returned to our new normal.

The seeding was going fine, and we had almost finished in the spring of 2005, when heading out into the farmyard one morning, I developed a problem walking and was unable to catch my breath. I called out for help, and soon Donna and Stephen arrived. Stephen, a trained first responder, grabbed the oxygen tank from his medical kit. To save time, Donna drove me to Regina General Hospital where I had been taken four short years earlier. An ambulance would have taken at least half an hour to come to the farm, but the nurses at the hospital told Donna she should have called one so that they could have started an IV and thoroughly reviewed my history before I arrived at the hospital. Lesson learned.

At the hospital, my stats were taken, and it was determined that my blood pressure had dropped, I needed more oxygen. I was nauseous and had fluid in my lungs. There was not a huge amount of heart damage, because my blockage was affecting the oxygenated side of the heart. After ten days, a second stent, and more medication, I stabilized and was sent home with oxygen, which I needed to use twenty-four hours a day for the next six months.

I had had another wakeup call, perhaps to again revaluate my lifestyle concerning diet, exercise, and work. Throughout the summer, my family doctor's, Dr. Stein and Dr. Claasen, and my cardiologist, Dr. Ma, monitored me closely.

That September, when I was back living in Calgary, I tried cardiac rehab at the Talisman Centre. Their physicians decided not to give me a stress test, because my heart was not contracting well enough for the walking. In October, another angiogram was performed at Foothills Hospital. The diagnosis was clear now: I had congestive heart failure. In November, another test confirmed my past diagnosis, which was that my left ventricular was mildly enlarged.

My quality of life, hopefully, depended on a change of diet, additional medications, and taking it easy. My new routine consisted of tests, appointments and medication adjustments, which continued into 2006.

"How is your state of mind, Doug?" asked Nurse Dana.

"I'm very worried about my family and what they will have to deal concerning the future of our businesses, but I'm not giving up. I just need a little more time. I think things will work out, but I'm resolved with my fate," I said. One must believe.

We spent the 2006 Christmas season in Earl Grey and Yorkton, where we enjoyed our grandchildren and had a great visit. Donna had arranged a cruise for my birthday, so we headed to Florida to board a cruise ship called The Miracle, a fitting name for my birthday and our future. The trip took us south to Panama and the Panama Canal. The seaway locks and the passage to the Pacific Ocean in Central America were magnificent. The magnitude of the Panama Canal building project, with its French and American history and the tremendous loss of life, was so overwhelming. Its capacity to move gigantic ocean liners through this human-made waterway was breathtaking. The story of the construction pre-World War II, its direct passage to Pearl Harbor, and its contribution to winning the war invigorated me. That feeling reminded me of my first visit to Pearl Harbor in Hawaii, in 1976 for our honeymoon, with the war ship masts rising out of the ocean, so historical.

The return trip along the coast, the side trips to lavish rainforests with wild monkeys climbing trees along the river, and the beautiful waters around Belize were breathtaking. Today's technology of global positioning, communications, and structural design play such a role in our pleasure and entertainment. The exposure to new places always makes me realize the small yet significant part we all play in improving the world. The trip empowered me to want to do more. But a cold I developed upon our return to Calgary soon became acute bronchitis, and I needed expert care once again. I was off to the hospital for another stay.

A week or so later, with the help of Nurse Dana and the hospital staff, I was back home and into my routine. The wellness program I attended twice a week at the Talisman Centre in Calgary was tiring and challenged me. I was accustomed to other people passing me on the walking track; however, one day a much older gentleman passed me with his walker. Well, what a reality check that was. This little incident evoked a terrible feeling of fragility and sense of my reality.

The rest of winter 2006 consisted of medical appointments, medicine changes, and the emotional ups and downs of good and bad days. I was always so grateful for the excellent care I received and Donna's endless effort and support for me. Driving me to all my appointments was a full-time job, never mind the waiting and never knowing how long before we could go home.

On April 4, 2006, Nurse Maureen measured my blood pressure at eighty-four over sixty and my weight at 193 pounds. I needed to keep taking deep breaths. In May 2006, after trying fish oil and all manner of experimental medicines, Dr. Tyler examined my candidacy for a pacemaker. He decided against it, determining it was too risky. For the rest of the summer, I stabilized and carried on with daily living. I am not sure why fall always triggered problems for me, but I was again back in the hospital for another two and a half week stay to regenerate me.

Wearing her red poppy on this frosty November 10, 2006, a full day before Remembrance Day, my nurse, Mary Smith-Martin, happened to mention that her mom had, like herself, been a nurse. When serving in England during wartime deprivation, her mom learned all manner of tried-and-true British home remedies. For example, birthing moms were treated to a glass of stout beer because the hops rushed in their mother's milk.

"My mother's name," she said proudly, "was Lieutenant Mary Smith. She volunteered overseas at the #24 Canadian Hospital in Surrey."

"I don't believe it," said Donna, bristling with excitement. The next morning, Donna appeared bright and chipper with her mom's wartime album, which featured photos of the brick building where her mom, Lieutenant Mildred Lanskail, also served at #24 Canadian Hospital in Surrey England during World War II.

"Recognize this woman?" Donna asked Nurse Martin when she arrived.

"Oh my gosh! Mom! That is my mom! Oh! Oh, wonderful!" Nurse Martin wept instantly.

"Your mother, Lieutenant Mary Smith, was my mother's bridesmaid on January 31, 1946 in England!"

"Yes!" exclaimed Nurse Martin. "Yes, my Mom told me she had carried a bouquet of rare wartime pink carnations for her best friend's wedding. I, too, visited that Roman Catholic chapel in Redhill, Surrey! Mom said her dear friend, Mildred, was a tremendous mentor to the anxious young nurses coming in."

"After the war," said Donna, "Mom volunteered with the Girl Guides of Canada as a ranger leader. When she died, the flag pole in front of the Guide office in Calgary was dedicated to her to honour her outstanding service."

A crackle over the intercom interrupted us. "A moment of silence," issued the voice of the hospital director, "on this November 11, Remembrance Day, for all Canadians who dedicated themselves to fight the brutal oppression of one of the worst bullies the world has ever known. We shall now listen to 'The Last Post,' a reading of the fourth verse of the 'Ode of Remembrance,' and bow our heads for two minutes of silence at precisely 11:00 a.m. After the service, wreaths will be laid at our local war memorials. On a personal note, I might add there will never be enough garlands to thank all who pooled their talents to rescue those so cruelly oppressed. Thank you to our patients, staff, volunteers, and visitors, whose family members and friends we honour today."

"It's a small, small world, after all," said Donna, giving a hug to the daughter of her mom's best friend. That Remembrance Day, I felt enormous gratitude for the rescue of my life by the Canadian medical professionals whose dedication never wavers, wartime or peacetime.

—

January 2007 found Donna and me enjoying warm weather and relaxing in our motor home in Palm Springs, after Stephen had driven it to Arizona for us. Donna had taken over the main driving from Las Vegas to Yuma, Palm Springs and Oceanside, California. Donna loves Vegas, so we stopped there for a few days. The weather helped my breathing, and Donna loved the sun.

In early March, I developed a bad cold and needed medical intervention. Fearing the worst, we flew home immediately. Going directly to the hospital, I was once again under the excellent care of Dr. Ma and his assistant, Nurse Maureen.

It was my sister, Donna, who first suggested we consider a possible heart transplant. Dr. Ma, however, was cautious because of the risks and after affects. It was also my sister who first talked about a private clinic in the United States. In her research, Donna learned about the necessary referrals and many requirements. After that, with our request, Nurse Maureen said she'd send a letter to the Alberta transplant team at Foothills Hospital.

In March 2007, we received great news! We learned that Dr. Isaac and her team would consider me for a heart transplant. The transplant team informed us of the many hurdles, such as needing time to complete the tests required for my approval. I was quickly taken under their care, and we soon learned about critical factors, such as age (I was 64), overall health, infection risks, family support, and proximity to the hospital both for the initial surgery and follow-ups well into the future. I soon saw the wisdom of living close to the front doors of the Foothills hospital.

In the middle of the pre-tests, I took a turn for the worse. My heart was failing. The testing and decision-making sped up. I had two weak teeth pulled, due to the possibility of future infection, and now in the hospital already, I began rigorous breathing exercises to strengthen my lungs.

On April 10, an internal cardiac device (ICD) was inserted under the flesh of my left shoulder, with wires inserted into my heart. Its job was to start my heart pumping, should it stop.

Thank goodness I did not foresee the trauma in store for me. Going for a test or procedure was scary. I never knew what was coming. For example, when I was told two teeth had to be removed to prevent infection, my stomach recoiled. But despite always fearing dentists, I agreed. "I can do this," I said to myself more than a few times. Well, the hospital dentist spent three hours with me but could not remove them. This meant two more agonizing hours the next day before those tough roots were extracted.

One day, in line at the x-ray clinic, a man in his sixties was waiting by himself for his first X-ray. I struck up a conversation with him, and he finally said, "I'm scared." I could so relate and felt sorry for him. Without immediate support, the unknown is scary.

The hospital stays got longer, and I got weaker. I noticed doctors talking to worried family members in the corner of the room, out of earshot. It felt disheartening. Suddenly my kidneys began to fail as well, and I needed dialysis. A procedure placed a tube in my neck that stuck out over my shoulder; it was freaky looking. On one of my required walks around the unit, I passed a young child who looked up at me, confused at the sight. It was also unnerving and stressful for my family to see me with all the cables, bells, and whistles that kept me alive.

A nurse asked me one day, "Do you have persistent pain?"

Although I was terribly uncomfortable much of the time, thanks to pain medication, the pain never persisted for long periods of time.

Soon, my heart gave up. It was too weak. The ICD emerging out from my shoulder was no longer of use, because starting my weak, failing heart was no longer the issue. The doctor's summoned my family. I was extremely sick and sedated, but I knew my family was with me. At that time, I had no concept of a Ventricular Assist Device (VAD) and its capacity to prolong my life while I waited for a heart transplant.

On May 3, 2007, with no other hope of surviving, I was wheeled into surgery to receive an internal heart pump. I was the first patient in Alberta to obtain an internal, rather than an external, heart pump, with five protruding tubes from my chest, four liquid drainage outlets, and one larger tube for connecting to the portable compressor type machine.

I remember little of the few days following my surgery. What a traumatizing sight for my family to see me post-surgery, decorated with all the cables, wires, and tubes that kept me alive. But I woke up, and I was alive! I had no complaints about the constant monitoring and draining of the tubes. I was content to be alive.

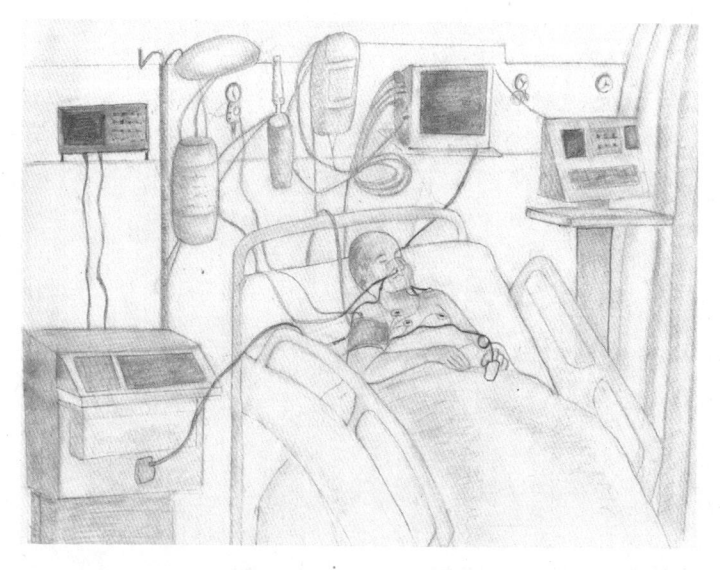

Earlier in this hospital stay, I became paranoid about catching pneumonia. I did not want to lay on my back for long periods of time. The staff was understanding and ensured I did regular walks, and they eventually provided an easy chair for me to sit up in, per my insistence. With no sign of pneumonia or other major setbacks, I continued to improve with each passing day. After at least two weeks without a shower, one was finally scheduled. Being hosed down reminded me of farmers washing their prize livestock at the cattle shows, with me sitting in the middle of a dingy concrete room and the attendant holding the shower hose and soap. But I had no complaints; it felt great to be clean.

—

"Dear Edna Brower," said Tommy Douglas. "Even as we view Doug's VAD implant this May 2007, you are among the committed humanitarians we thank." Rising to his feet, he addressed a beloved former Saskatchewan school teacher, a woman he considered a brilliant behind-the-scenes politician.

"You, Edna, backed my initiative to introduce provincial hospital insurance in 1947," he said, adjusting eyeglasses he no longer needed now that his vision was perfect. "When your husband walked into my Saskatchewan office to endorse my recommendation for universal healthcare, I knew who had encouraged him. I thank you, Edna Brower Diefenbaker, for all you did for Canada."

"Medicare was born in Saskatchewan on July 1, 1962," said Edna, "and is of great pride to all prairie Canadians who fought hard for it."

I have been told that Edna Brower, John Diefenbaker's first wife who passed away at a young age, was said to be a descendent of William Brewster; however, this has yet to be verified.

—

The new program of an internal valve and pump, which in 2007 sat in a satchel on wheels, was supervised by Nurse Anita, or Super Nurse, as I called her. She was one of the most capable medical professionals I met throughout my ordeal. An expert in her field, she exercised the utmost defense against infection and instructed Donna on dressing changes and the critical importance of sterile conditions around me. Donna was an A+ student. When I was so sick and certain I could not walk another step, Anita pushed my limits even further. Our family much appreciated all the devoted transplant staff who worked tirelessly to ensure the VAD's success in keeping me alive. Looking at the machine and the attached hose sunk into my flesh put things into perspective for me.

As the days passed, Nurse Anita had me venturing out and about to the Market Mall, or down by the river, regardless of my fears and refusals. Exhausted after a walk with Donna and Cathy on the hospital grounds, I would hear Anita's suggestion, "Let's just go a little further, Doug." Up to the next corner, just across to Tim Hortons, or just around the next bend. Nurse Anita also trained Donna and Cathy on the operation of the complex VAD pumping device. Donna learned to charge and change the battery at home and to handle any malfunction of it. Donna joked with Anita that she would only unplug me in the morning to make coffee, because the electrical outlet would only allow for one of us at a time.

Donna studied the operating manual and learned to manually create the necessary air, in case of an emergency with no power. At that time, at my lowest ebb, I was so sick and so emotionally beyond caring that I dreaded the visit of this guiding angel who would not go away or take no for an answer. The time finally had arrived to leave the hospital with my VAD, the discharge and transition preparations began. However, a problem arose immediately upon leaving the hospital.

We had been living in our motor home at the Balzac farm the past summer and gone south for the cold winter months when my health took a turn for the worse. Our friends Rick and Margo kindly brought our unit back to Calgary, but it was still too cold to live in. We had not yet found permanent housing. So, here we were, ready to go home, but where was that? It was late spring and the Calgary West Campground, near the hospital, was opening. Happily, they had a spot for our mobile home. Rick again set up our unit for us, which was a lifesaver for Donna. We were extremely grateful for the much-needed help.

Worried that the first internal VAD outpatient was moving into a motor home, Nurse Anita called Fleetwood, the manufacturer, about the electrical system and was assured our unit more than met the required electrical standards (even a generator. I would have all the backup advantages I'd have at the farm or hospital.

Another issue was where I could sit in our car. Donna applied to Transport Canada for an exemption to turn off the front passenger seat airbag. Neither of us wanted an airbag exploding onto my chest in case of an accident. Even though the transplant unit supported our request, it was refused. I would have to sit in the back seat while the car was in motion.

Dr. Burden, a founding member of the transplant program, escorted me out on discharge day. I noticed the beaming smiles and nods of several men and women. After he congratulated me on having earned my discharge papers, Donna quietly explained to me that the nods and smiles were donors to the Libin Foundation here on a tour.

Here we were on May 7, 2007, and I was the first Albertan to receive an internal heart pump while waiting for an organ transplant. Until that moment, I had no idea of the vast number of caring people responsible for my recovery period. Donna explained that in 2003, the Alvin and Mona Libin Foundation presented the largest one-time donation to the Alberta Health Services and the University of Calgary, funding a world-class institute for healthcare research, education, and patient care. I could feel my stronger heartbeat in my stapled chest as I looked their way. Once again, I felt so lucky.

I was only too happy to be a poster boy for the Libin Cardiovascular Institute of Alberta. They rightfully celebrated their wonderful lifesaving VAD program and its benefits for so many patients. A big plus was that patients could now recover in the comfort and familiarity of their homes. I was a proud first recruit. The Calgary Herald photographed and wrote up the story, and soon enough, my old friend from my Commodore days, Lloyd, recognized me and called a few weeks later.

Michelle Lang, Calgary Herald journalist, planned to continue following my VAD progress. Tragically, she lost her life while reporting on Canadian soldiers in Afghanistan a few months later. When I heard the terrible news, I thought of how reality changes. In Ms. Lang's case, it was reality at its worst.

CALGARY HERALD Friday, May 25, 2007

HEALTH

Homecoming eases transplant anxiety

MICHELLE LANG
CALGARY HERALD

Calgary surgeons have launched a new phase in a program that installs mechanical heart pumps in cardiac patients, allowing a local man with the device to return home Thursday.

Albertans who receive one of the heart pumps have traditionally remained in hospital, but the local team believes some patients will recover more quickly at home as they prepare for a heart transplant.

"This will be tremendous for me," said Calgarian Doug Brewster, 64, the first patient in Alberta to leave hospital with the device. "It will build me up."

Arranging the homecoming involved complex planning to ensure there is no disruption to power, which could jeopardize the mechanical pump.

Brewster has backup batteries that his family has been trained to change every hour and a half.

His home has a generator that will provide power if electricity goes out in his neighbourhood. And, if all else fails, his wife and daughter have been schooled in using a manual pump to keep his device working.

"I'm totally comfortable with it," said Donna Brewster, his wife.

The decision to send a patient with the device home comes nine months after the Calgary Health Region first began a mechanical heart pump program last September.

SEE HEART, PAGE B4

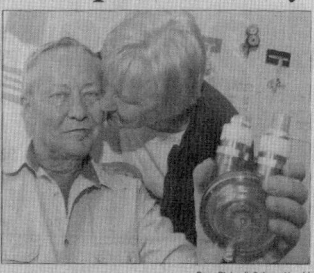

Dean Bicknell, Calgary Herald
Doug Brewster is the first patient in Alberta being sent home with a heart pump to await his call for a live-organ transplant.

FROM B1

HEART: Many recover quicker at home, says doctor

The pumps — which are called ventricular assist devices — keep patients like Brewster alive until they can undergo a heart transplant.

Ventricular assist devices can also allow some patients with temporary cardiac conditions to recover.

In rare cases, the pumps can sustain patients for the rest of their lives, although that rarely occurs in Canada.

Allowing a patient with the device to leave hospital while they wait for transplant marks a new chapter in the Calgary program.

Physicians estimate between six and eight Calgary patients a year could be able to go home.

Some other Canadian cities also allow patients with ventricular assist devices to go home while waiting for a transplant.

Dr. Debra Isaac, director of cardiac transplant for CHR, said recovering at home helps patients like Brewster prepare for transplant surgery by becoming stronger through daily activities.

"Hospitals aren't good places to be if you don't need to be there," said Isaac.

Indeed, Dr. Paul Fedak, a Calgary cardiac surgeon who has worked with the heart pumps in other cities, said he has seen many patients recover more quickly at home.

"They are a person in the community, not a patient," said Fedak.

Allowing Brewster the freedom to leave hospital involved detailed contingency planning. For instance, Anita Hadley, a nurse who co-ordinates the heart pump initiative, had to train area paramedics in treating Brewster, should anything go wrong.

For his part, Brewster has to remember to be continually hooked up to a power source. He can plug into power outlets in the walls at his home and even the lighter in his car. But physicians encourage him to use the batteries on a portable power pack so he can move around freely, fostering a speedy recovery.

"I'm looking forward to getting back to normal," said Brewster.

MLANG@THEHERALD.CANWEST.COM

LUCKY ME

DONNA AND I FINALLY SETTLED INTO A routine and adjusted to our new-found friend, the VAD. I soon learned to wheel around a thirty-five-pound suitcase, which produced a chugging noise, in a 230 square foot living space. Our motor home had a couple of six-inch floor risers and front doorsteps going to the ground, but I navigated these with Donna's help. The grocery store cart was a perfect place to put the machine, with me pushing it along, gathering groceries. Restaurant tables were high enough so that the device could tuck underneath it on our supper outings. However, the nightly jaunts to the bathroom meant disturbing Donna so she could help move the cart. A caregiver's sacrifice never ends. The machine noise, and my reliance on it, was not an issue for us. Perhaps that is because we were farm people, after all, and were used to being around machinery. I was often asked about the mental effects of my dependency on the VAD, and I would reply that I was getting stronger, which was excellent, and we were settling into everyday life with no thoughts of any quick and immediate changes.

We had no idea when a donor heart would become available, only that I was on the list, packed, and ready to leave on a moment's notice. Donna's cell phone was always to be free to receive a call from the hospital. We were told we must have a process of notification for our family without jeopardizing the communication line with the hospital. There are many stories of failed transplants because recipients could not be reached or were unavailable for surgery. We were determined not to be one of these stories. Issues of compatibility and there even being a heart available are further testament that there were no guarantees with this process. Because of transplant costs in the USA and no overall government medical coverage, we learned that many patients live with a VAD, not as a temporary measure, but all their remaining lives.

—

On June 6, we went for a walk along the river after supper. Later, we were getting ready for bed after watching the CBC News. Then the call came. It was our time!

Per our instructions, we made our one permitted phone call to Cathy, she made three more calls, and those three were to fan out from there. We missed Anita's first alert because we were on the phone, but she dialed again and again. Overwhelmed with emotion, we were told an ambulance would pick us up in twenty minutes.

We arrived at the commercial area of the airport by ambulance at 9:30 p.m. on June 10, 2007. A fully staffed plane sat between two hangers, awaiting our arrival. I was soon secured onto a mobile gurney, and two medical specialists guided me into position in the orange twin-engine air ambulance. Donna would fly with us, as she was the one who could supervise the VAD that was keeping me alive. We taxied on the runway and were quickly airborne; however, our flight hit turbulent weather, which altered our flight time. Just another hitch in a long journey.

Donna and I have an unspoken policy for times like these. We assume a calm, business-as-usual manner for the sake of each other.

Donna did the checklist. In my carry case was a list of my medications. Donna organized that six weeks ago when my VAD was inserted. Included were the immunosuppressant that shut down my body's natural immune responses that could damage my new heart. There were antibiotics, antivirals, and fungicides that fended off infection when the immunosuppressant disabled my natural immune responses.

I signed the official consent form. Donna witnessed it.

The medic explained, "As soon as we land at the old downtown Edmonton airport, an ambulance will take you to the University of Alberta Hospital and from there, up to ICU. Your wife will remain with you to look after your VAD. Then you will have blood work, an ECG, and a computer chest x-ray, which will appear on the overhead screen in the OR. Finally, an intravenous line will be installed to measure the pressure in your lung arteries. All this," the knowledgeable medic continued, "will take only minutes. Then your anesthetist will appear."

"Thank you." I said, glad to know the agenda.

Cathy drove Donna's car to Edmonton to sit with her for the long night ahead. Stephen, Bettyanne, and sister Donna were on the road from Saskatchewan. They arrived at 5:00 a.m. I smiled at Donna and Cathy, possibly for the last time.

"Once you are asleep, Mr. Brewster, your surgeon, will insert a windpipe tube, which will be connected to a respirator to support your breathing during your surgery. Next," she continued in reassuring tones, "a tube will be placed in your stomach to stop liquid and air from collecting there, so you will not feel sick when you awaken. Another tube will be inserted into the bladder, too.

Terrified, I just nodded my head. I was grateful for the heated hospital blankets being placed over me in the cold operating room. I noticed what appeared to be an ancient stitching machine above the bed, all ready, I expected, to sew my chest shut after surgery.

The anesthesiologist was concerned that my rather small throat opening might not accommodate all the tubes required. Just before I panicked, he said, "Do not you worry, Doug. I have a device to expand the airway." I envisioned a large mechanical apparatus from our shop, which really did not ease my fears. Finally, I was asked to count backward and was transferred onto life support. I knew that nothing was assured.

It was an awfully long day for Donna and the family. I was in surgery, from 4.30 a.m. until 5:00 p.m., during which time they received absolutely no information about my surgery progress. They didn't receive any information until later that night because, I expect, until the surgery is complete and the new heart working, progress and success cannot be determined.

—

I woke up in ICU, my transplant surgery appearing to have been a success. My emotions were intense and raw, and I began to reacquaint myself with the world. My body was in pain, my arms were tied to the side of the bed rails, and tubes were still in my throat, so I could not speak. It was incredibly stressful, especially with my fear of closed-in surroundings. It brought me back to thoughts of my father's grain bin, back in 1952.

I could hear and see my nurses watching over my movements and the wall of monitors around my bed. All of a sudden, there were other people in the room, and everyone appeared to leave. I panicked, I had to get my arms free. I somehow finally got the attention of someone who came and said my nurse had gone on lunch break and nothing could be done until she returned. After what seemed like eternity, she returned. Obviously seeing my problem she was still not about to act or call her supervisor.

I am not sure the commotion I caused was all my fault, but after communicating on a writing board, and my nurse finally placed a call to Dr. Isaac in Calgary. My arms were then released, and I was thankful to Dr. Isaac, and began to relax. I expect that call saved my life. I heard the next day a special team with their equipment were placed on standby all that night, in case of resulting complications.

The ICU was a huge room with a lot of patients. We were next to each other but separated by curtains. I initially recovered quickly and within forty-eight hours was sitting on the edge of my bed. I even took a few short walks!

The next couple of days I had problems with my IV and bleeding. On June 18, there was a realization that my own electrical charge was weak, practically non-existent, mainly due to scar tissue left around the heart cavity from the original heart removal. Talk of a pacemaker began, and I remember asking why they had not left the pocket on my right shoulder for the defoliator with a zipper for possible future use. Later that day, when I was heavily medicated, I asked Donna to call Rick about buying a floating barge and a wet suit for him to put his mini-excavator hoe on to mine gold underwater.

The medications were continuously changed and caused swelling in my body, particularly in my legs, which became huge. I had to wear special elastic stockings, and I looked great. The medications also affected the diabetes I had developed.

On June 19, while taking a short walk, I thought about Nurse Anita and commented to Donna and Cathy that even with my walker, she would be saying, "Let's just go a little further to that corner." On June 20, Dr. Burden, one of the founding members of the transplant program, and his staff visited me. On June 23, Donna commented, to the staff on duty, "Never in our wildest dreams would we be where we are today. Thank you. We have so much to be thankful for." On June 25, Dr. James, my surgeon, explained the need for a pacemaker and said he would transfer me back to Calgary. This pleased Donna because she needed her security back home with our transplant team.

On June 27, around 4:00 p.m., I was on my way to the Foothills Hospital in Calgary. It was quite a fiasco, with neither hospital fully prepared. As most people know, being released from or admitted to a hospital is usually a time-consuming and disruptive process, and this one was no exception. I arrived in Calgary with no medications to accompany me. The transplant team had to scramble to order the thirty plus tablets that I was taking at the time.

Donna was relieved to be home after living in a hotel room for three weeks. However, two eviction notices were on the motor home upon our arrival. She had arranged for our absence, but with staff changes, someone forgot, but the matter was cleared up immediately. I truly do not know how Donna managed through this difficult time. I know all our family gave her great support, but I am sure she felt overwhelmed and alone with me being no help, unhappy, and complaining. Her tremendous and courageous support got me, us, through this challenging but unbelievable gift.

A problem occurred the morning of June 28 when Dr. Jeff tried to open a plugged vein in my neck. Opening it would allow for the heart biopsy to go through my neck rather than my groin, which was much easier and less stressful. Dr. Warnica and another doctor worked diligently for an hour and a half trying to blast the vein open, but unfortunately, they had no luck. The next day, a pacemaker was placed on my right shoulder by Dr. Child who was experiencing difficulties. I awoke to the frustrated doctor using profanity and calling the transplant team, saying, "There is too much blood in the shoulder area to complete the pacemaker implant." It was impossible to complete. I expect he was told there was no alternative and the job had to be completed. Fortunately, he was successful.

Shortly thereafter, Canada Day was upon us, and Donna was able to take me for my first wheelchair outing. A couple of days later, I had the first of many biopsies to check for heart rejection.

—

As time continued, I began to feel better and have free time on my hands! I began to evaluate the general concept of few people wanting or thinking they should pay taxes (especially income taxes). It seemed ironic that I, a professional accountant, trained to help people find ways to minimize their income tax payments under our Canadian Income Tax Act. I grappled with the perception that my profession was often accused of helping clients to avoid paying their fair share.

It became much clearer to me what government tax revenue does for the average citizen. I cannot even imagine the cost for my care over the past years, or what the future cost would be. Thank goodness for our Canadian medical system. It may not be the perfect system, with problems such as long wait times and not enough doctors, nurses and support staff, but it is still superior to a private-pay system. Taxes also pay for roads, streets, community centres, and schools.

One day, as my well-used brittle plastic cables were untangled and replaced by two attendants, I overheard the attendants discussing their overtime hours in their workplace. They said working overtime was not worth it because a good portion of the payment went to the government for taxes. I tried to explain that government deductions are their own money. For example, $100.00 in wages less $15.00 tax equals $85.00 for the employee now and $15.00 held for their benefit to pay for their necessities, such as roads, medical care, schools, and police services.

I began physio which would continue for some time. My physiotherapist, Ruth, a seasoned practitioner, visited to discuss firming up the connective tissue of my new heart and improving my muscle coordination after so many months of virtually no exercise. I had to learn balance again, as well as going up and down stairs. It is amazing how inactivity affects our bodies so quickly.

"Let's proceed," she said, moments later in the gym, and four words sprang unbidden to my lips.

"With all due caution."

"You'll stop at every station," she said, pointing at each stop around the gym. "The last one is home."

"Ready," I replied. "I want to earn my way home."

An hour later, exhausted beyond human comprehension, I sank into my bed and imagined myself driving my truck home for dinner.

—

I had slept all afternoon, tube free, although now I swallowed a cocktail of thirty pills a day, starting with the mandatory blood thinners. Tilting her head to one side, as she often did when initiating a conversation, my daughter Cathy commented, "I've read, Dad, that kids tend to compensate for their parents unlived goals. The article suggests that children feel an obligation to live out their parents' unlived dreams."

Hmmm, I considered. "That's sad," I said. "So, what are you thinking, Cathy?"

"Well, you've had three heart attacks, Dad. You've been so busy."

"Go on," I said, listening to my thoughtful youngest.

"Well, I may be completely wrong, Dad, but I wonder if you felt compelled to take care of Grandpa Ken's fear of poverty. Maybe at some point you decided that none of your kids would worry about the cost of fuel to go see a movie with friends."

"So, you think I went overboard in the work department, Cathy?" Taking a sip of her Tim Horton's double-double, Cathy waited. "So, what new venture do you think I might take, Cathy?

"The retirement scene maybe? Fun and travel with Mom?" Cathy replied.

Shortly after Cathy left for the afternoon, a thoughtful hospital volunteer wheeled my bed table in place, stabilized my laptop, and got me plugged in. I reflected upon Cathy's respectfully submitted ideas. I considered that yes, indeed, I could insert a few lazy days on the British Columbia coast, at our favourite spot, Pacific Sands in Tofino, with grandsons Colton and Camden. Maybe I'd attend a lecture or two at Calgary's Kerby Centre, or just simply relax. Maybe, I took on Dad's fears, but they were long gone. If I had been afraid of poverty, I was not anymore. But sure, Cathy was right. Slowing down was fine. I was all in.

—

On July 8, Stephen and Amanda visited. Stephen asked about a legal problem with Barry, a fellow farmer. He said our lawyer wondered if we wanted to change our decision on how they are to proceed with the matter. I said, "Tell him this senior partner knows he has had a change of heart, but his mind is still the same. Forge ahead."

On July 9, I was discharged from the hospital. Donna and I were so happy to be heading to our newly purchased home near the hospital. There would be many trips to and from, daily and then twice a week for the first year. It is hard to put into words how we felt after such a long and treacherous road, filled with fear and uncertainties for so many months, was now ending with us sitting in our own home. A few days after getting home, Donna transported a purchase using a Home Depot van. I climbed into the driver's seat, and Donna took a photo. I sent it to the transplant team and said, "I found a job!" I'm sure they must have wondered what was going on.

"You've got mail, honey," said Donna. Her 1970s music was playing, and the two of us were finally enjoying some much-anticipated quiet time. Soon enough we would begin our daily routine of hospital visits for biopsies, medication changes, physio, and checkups. There were many get-well cards from friends and family to whom we were so thankful for their support and concern. I had also received a letter from an unknown address, which Donna read to me. It was as follows:

July 2007,

Dear Mr. Brewster,

My wife, Jasna, and I, our two daughters, Berit and Clara, and my son, Helmut, join in wishing you a strong recovery from your magnificent heart transplant. Because you've generously shared your heartfelt advice and counsel too many people like us in your government-appointed work as a Farmland Security Board member, my family and I agree that it's your turn to receive, Mr. Brewster.

Once again, our family thanks you for your advice at the Farmland Security Board meetings, where you and the committee met with us twice. Even though each meeting lasted four hours, you stayed with us all to help us with our farm's financial crisis. The help you provided, for working through our problems after three years of drought with our creditors across the table in an organized manner, made the difference in us continuing to farm and providing for our family.

Perhaps today is a good day to share with you the name of my father, the German gentleman your father hired to stook hay a long, long time ago. Jasna and I named our son, Helmut, after him. Post war, my dad, an objector to Hitler, hid himself up to his neck inside a rail oil tanker headed out of East Germany. Starving and alone, it was the farmers along the back roads who fed Dad in exchange for his work. Dad once told me about you and the artful way you showed him to stand up four oat stooks to dry, one leaning slightly upon the other (just like a family does), not too much, though, or one could fall. Dad told me that even back then, your father, Kenneth Brewster, knew you were a good teacher. Although aged now, our father sends you greetings and his thanks. He said that first your father, and then you, have cared for our family so well.

The Schneider Family

BEGINNING AGAIN

LATE AUGUST 2007, I WAS READY TO venture out of Calgary. Soon, with the transplant team approval, Donna drove our motor home to Banff for a weekend. It was great getting out in nature; I so needed it. Next airline tickets were booked, and we were off to the farm in Saskatchewan to visit family and friends and see how harvest was proceeding.

The first two years post-transplant were both joyous and frustrating. Despite our worries and hopes for our future, we stayed positive. Amanda and Stephen's wedding took place on December 27, 2008. They had a large celebration with lots of friends and family, some of whom travelled long distances to join them. It was great to catch up with everyone. Stephen and Amanda made sure each wedding table had heart donor cards and lapel pins for everyone to promote the transplant program and its benefits. In addition, whenever Donna and I were asked to talk with a patient about transplant, we were always happy to do so. We also provided our basement suite to patients and their families from out of town, so they were able to stay in a home environment close to the hospital. The couples were so appreciative of our care and support, helping them through the recovery period.

We wrote an anonymous letter to the donor family of my heart, thanking them and trying to convey the depth of our gratitude for their gift to me. We appreciated receiving a beautiful letter back from this courageous family. Along with the many trips to the hospital and changes to my medications, I had been dealing with post-transplant problems, such as osteoporosis in my back, diabetes, and the side effects of taking thirty pills a day. Still, I was optimistic.

A transplant patient is said to have no feeling around their new heart, as the nerve endings have been severed. However, during the pre-Christmas season in 2009, I had a feeling something was wrong. I was experiencing some pressure in my chest after supper one evening, so I was taken to emergency by ambulance. Again, I need an immediate angiogram, which determined there was another blockage. A stent was inserted, but unfortunately there was some damage done to my new heart.

After excellent care, I returned home from the hospital ten days later. But once again, I was depressed and unnerved that this was now happening with my new heart. *Where do we go from here?* Donna and I wondered. The future seemed uncertain, but as the old saying goes, you put one foot in front of the other and move yourself ahead.

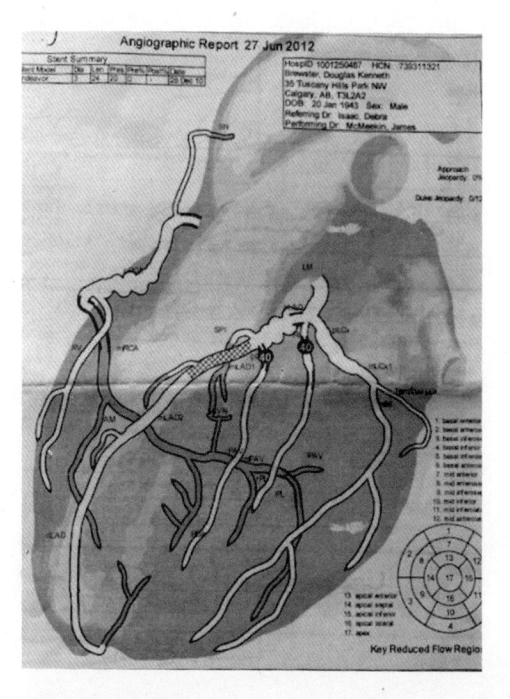

At our first transplant clinic assessment, we heard grave concern with the damage to my new heart. Donna and I left the appointment discouraged, with many concerns and questions. Had all we had gone through the last couple of years been worth it? Being proactive people, Donna and I asked for the help of Dr. Isaac and Bonnie, another superb nurse. Fortunately, I received another angiogram and further tests, with another complete review of my situation by the transplant team.

Adjustments to medications and lifestyle changes encouraged a brighter outlook on our future. Donna and I have found that a patient, and sometimes their family, must advocate for themselves. This does not mean second-guessing our medical professional; it means respectfully letting very busy professionals know our thoughts and concerns. The medical system expected me to be honest, to provide forthcoming information and to be clear about my concerns.

—

Stephen had been managing and running the rental and sales operation of Douglas Industries Ltd. in Earl Grey. It was officially shut down in 2010 and taken over by Stephen and Amanda's farm operation, Brewster Ag Industries Ltd. I sent a letter to the community and past customers expressing our thanks for their support over the past years. It was a long list, as Douglas Industries Ltd. began in 1972 (along with Brewster Trailer Manufacturing) and had been the backbone for many operations and activities since that time. Eventually, the business was deregistered, and its final tax return was submitted. This company was born, lived, and died as does a human life.

Over the next few years, our life settled into a routine. Donna was busy in the spring months volunteering her income tax skills at the Kerby Senior Centre. She still enjoys doing tax for regulars who ask for her and loves the treats she is given. Donna never turns down chocolate. She enjoys her house and yard, in which she shoulders most of the work. Summers are spent in Saskatchewan at our little farmhouse on Stephen and Amanda's farm. We still take short trips, mostly by air, and continue to enjoy the west coast of Victoria and California. We love the water.

In 2015, I received a call from a fellow by the name of Steve who said he had something from Brewster Trailer Manufacturing he wanted to return. He had carried around our labeled stapler for twenty-five years and wanted to give it back to us. Many years ago, Steve and a buddy of his from Ontario, had gone walking door to door, looking to find work. They stopped in at Brewster Trailer Manufacturing in Calgary. I hired both nice young men who, although not experienced, learned quickly and helped us transition the operation from Calgary to Balzac. Steve was a loyal and devoted worker for several of my future businesses and helped us move and get settled in Saskatchewan in 1985. He continues to live in Calgary and earned his Chartered Accountants' certification. He got married and now owns his own technology payroll company. He showed us his company facility, with its high-tech security system. We had lunch with Steve and his wife and reminisced about the old days, telling him how proud we were of his success.

I began online auction bidding and buying the odd piece of equipment for Stephen's farming operation. It is a great hobby. With time on my hands during the winter months in Calgary, I began to bid at Graham's Auction of industrial sales but soon bid on items from Grand Prairie, Alberta to Minneapolis, USA. Today, an auction can have internet bidders from miles away who visit from their computer screens. From my home office, I bid at Larry's (Graham Auctions) and Ritchie Brothers in Edmonton and Saskatoon. Anyone who has stood for hours in the winter at an outdoor auction understands the advantage of participating from the comfort of home, no matter the weather conditions.

Larry's old saying still holds, even in this new millennium. "There's smart and not-so-smart bidding. Just as auctioneering is an art, so is bidding." This is true. For example, an innocent buyer will pay $5,000.00 to Birks for a well-cut diamond ring for his love. A savvy buyer, on the other hand, will bid $1,500.00 at a barn auction for the same ring and then treat his sweetheart to a holiday in Maui with the savings. The key is knowing the condition and true value of the item you are trying to negotiate a price on; however, internet bidding makes this much more difficult.

Although healthy, my mobility is limited. I continue to run equipment, though, and oddly enough, I believe the bouncing movement helps my overall body system. I have developed a large hernia, though, from the surgeries, and deal with it daily, wearing a special belt around my waist.

I enjoy summers at the farm. I love nature. The Earl Grey area has an abundance of wildlife and natural trees and grass, with potholes full of water. It not uncommon to see a mother moose and her calf in the fields, a bald eagle flying overhead, a fox running, or white tail deer eating on the combined field at dusk in a single day. The surrounding Earl Grey area is also close to Regina, Last Mountain Lake, and the Qu'Appelle Valley, where they are proud of their vegetable gardens and host their summer and fall markets. The lake and river system attract many tourists and fishermen to the area with its scenery, boating, and great fishing for pickerel and north pike. Craven's Big Valley Jamboree is a world class music festival that we attended in the 1970s, with performances by musicians such as Willie Nelson and Kenny Rogers. In those days, Father Larry, a Catholic priest, ran the show with much controversy. This festival continues to be great summer entertainment for the prairie people.

On returning to the farm in 1985, the barn had greatly deteriorated and desperately needed repair, but with the roof structure only being able to support light cedar shingles and repair costs so high, it stood empty for another twenty-seven years. In 2012, while improving and enlarging Stephen and Amanda's yard, I dismantled the barn using our D6M dozer, saving only a few items to reuse.

Barn 2012

The barn was fifty feet long by sixty feet wide, with a twenty-four-foot-wide centre isle and two eighteen-foot-wide side portions, all fifty feet long. The main centre portion was designed for Mr. Pinkney's prize riding and work horses, with high mangers and harness holders at the end of each stall. The centre isle and the stall floor were cement, which were easy for cleaning, with lots of room to harness and groom the horses. Dad remodeled the stalls by lowering the mangers so three of the mature cows could be tied in each stall, rather than the original two horses along with some stalls being made into pens with poplar pole gates.

In the early 1950s, Dad made good use of the tall, straight field poplars scattered in bushes across the land. One time, he found a few branches that were just the right size for the job. He chopped them down by hand, cleaned off the branches (usually the bark peeled off), and made them ready for loading or dragging home. Lumber was costly and not readily available, so grain bin skids and bracing, gates, cattle feeders, tripods, and hundreds of fence posts were treated with bluestone. Poplar wood is soft and does not take a lot of wear sliding along the ground. It was amazing the years we would get from those skids on our stone boat, pulled by a single horse, used every day in the winter.

In 1950, the south side portion was pretty much an open area with a dirt floor. Dad, again using poplar poles cut from our poplar bluffs, built a large pen, feeding lane, and manger against the centre divider wall with doors opening from the hay loft above for straw to drop. In the fall, the spring calves reached six months old and would be weaned from their mothers in this pen, where they would then be fed over winter. When the calves were first taken off their mothers, they needed water brought to them for the first month, so they wouldn't return to their mothers for milk before graduating to the dugout with the rest of the herd. Dad dug a hole next to the calf pen and placed a forty-five-gallon barrel to store water and keep it from freezing. The water was brought in with another barrel placed on our horse drawn stone boat. The driver, after filling the barrel with a pail at the dugout, would carefully make his way back to the barn, trying not to get too wet before emptying the water into the storage container. There was no way of water not splashing out over everything with the rough ground and frozen manure. Dad closed the east end off with wire and poplar poles for chickens.

The north side had been built for milking cows, with a front feeding ally and manger next to the centre partition, then room for the milk cow in front of the manger and a narrow back alley behind the cow for access to milk. This was also where we'd bed the animals. We would milk a cow for fresh milk in this area, until Dad decided his time was better spent grain farming, following which the area was converted into grain storage.

These barns have large openings with a sliding door to the hay loft high in the front of the structure. This door was for mechanically moving feed, such as oat sheaves, into the hayloft for future use. In 1949, Dad refurbished the hay slings and rail with a new rope we used for a few years until square bales replaced the binder and stoking sheaves, rendering the moving system unusable due to the heavy bale weight from the baler, which compacted the oat straw rather than the grouping of grain stems with twine wrapped around it. The first balers used wire to secure grain stems and were much bigger and heavier.

In 1952, I remember Dad forking the sheaves from the field stooks onto the hay rack, stretching the slings across the entire rack every twenty-inches of sheaves with a full load of approximately ten feet, thus requiring four slings per load. The ride home on top of the load of sheaves was great fun for us kids, as Dad would hold the horse reins, sitting beside us, and steer them home while watching for bumps and holes, so as not to tip our load. Once in front of the barn, the large high loft door was manually lowered by moving counterweights that hung from ropes, which lowered the large loft door. Then a trolley was rolled along the rail out through the door opening with two thick ropes hanging down over the centre of our full load. One end was attached to the team of horses for pulling (having now been unhooked from the load) and the other attached to the ends of the top sling on the hay rack. As the horses pulled on the rope, the load sling came together making a big round circle of sheaves that went up to the top, sliding into the barn on the roof track and dropping its load, ready for the next bundle. Because I was involved and got to see lots of sheaves and bales being hand thrown and stacked in our loft over the years, I have continually marvelled at this early mechanized unloading system that saved so much hard work and time.

In the late 1950s, outside corrals with their own shelters came in, saving labour and allowing for large herds. Thus, these large cattle barns virtually became a thing of the past. For many hours I would help Dad with a spade and crowbar, digging four feet holes every ten feet around the outside of our new corral placing used railway ties in each hole and tamping the dirt and rocks around the post. Dad was particular about straight posts and placing only four inches of dirt in the hole around the post with a small row of rocks and tamping, then repeating the exercise until the hole was filled high enough around the post that water would run away and not rot the post. Poplar slabs from northern Saskatchewan were secured to the railway ties and nailed close together, giving protection from the weather and securing the livestock. The young animals or sick livestock still needed better housing from the elements and stayed in the barn. However, with the majority being kept outside, the chores were much easier. Dad continued to maintain the barn well, and I helped replace areas of the cedar shingles, which is a slow and tedious job.

Before long, with some minor renovations, the barn was converted into grain storage. Grain storage was usually in small wood structures that Dad had also constructed with his hand tools and my help, as premade metal round bins, grain augers, and truck hoists were just becoming available. Grain shovelling by hand was still the way to move grain from the truck to a storage bin in the early 1950s. The barn was good storage but made handling the grain difficult. It was hard work to shovel the grain straight into the barn out of the truck, and even with an auger moving it from the truck, lots of grain shovelling was necessary to move the grain into the stalls and mangers. The average person with a grain shovel or scoop can move grain ten to fifteen feet, so it was not uncommon to move the same grain at least six times before it left the yard.

In the years after I had left, Dad removed some partitions and made holes in the roof and hay-loft floor. The longer augers helped when filling it in harvest. I expect this was one of the reasons I hated shovelling and would design equipment to help or convince someone else to do the job.

—

It was difficult knowing the barn would be gone forever, but as a realist I understood it was time. The old cement was buried, along with many hand-picked field rocks used for the foundation. Donna, Colton, and Camden spent hours picking up the many old nails to prevent future tire flats. Now, new metal hopper bins sit where our barn had towered. I guess that is called progress.

In the cleanup, we were lucky enough to find a board from the barn peak with the initials "D.W." etched in the wood. Further research determined the initial belonged to Dave White, a local farm pioneer and carpenter who chiseled his initials into the wood with his framing hammer. We farmed his old homestead quarter and were great friends with his brother, Harry, who lived across from Fosterdale School. As a kid, I spent time at Harry's farm. We would have lunch in his kitchen, with little chicks chirping and penned off in the corner and all his kitchen utensils and food items stored on his kitchen table. Dad purchased Dave's home quarter and did a lot to clear the bush and old yard after I left the farm in 1964. Although, I do not remember seeing his actual yard on top of the knoll, I know its location and can imagine another pioneer's yard of the past.

NEW VENTURES AND FAMILY

IN THE WINTER OF 2013, I DECIDED it was time to do something on my own while continuing to help improve and aid in farm activities. My retirement project would be to sub-divide forty acres, zone, and develop a campground.

A quarter section we had worked and purchased from the Binnie family, just south of our yard, had an old single room house where Jim Binnie lived until the 1950s. It had twenty acres of natural bush and prairie, with many old trails where the original rural trail occurred. Back on one of the trails were the remnants of the original pioneer's yard. The story goes that those pioneers, by the name of Handyside, were gifted a lot in Regina because their child was the first baby born in Regina; however, they could not maintain the property tax and lost the lot. It was a perfect spot with good road access, so I proceeded with my new part-time venture.

We named the new campground Green Hills after a big open-pit coal mining operation in Crowsnest Pass, between British Columbia and Alberta, near a beautiful town called Elkford where

Cathy lives and works. The land preparation, campsites, building, and landscaping have been on-going for me and my equipment, with the much-appreciated help from Stephen, Amanda, George, Dennis, and as always, Donna. Also, with Bettyanne sending trees from northern Saskatchewan and Cathy clearing walking trails, it continues to be an ongoing family venture.

There have been many challenges for our new campground, such as four years of record moisture flooding a good portion of our beautiful landscape. Bugs came next and they stripped the leaves bare during our summer months for two consecutive years. Now the water holes are drying up and beavers are moving in pursuit of water; they are cutting down our live trees. However, we continue to plant trees to improve and enjoy the summer months on our own private campground.

—

In 2014, the Canadian Professional Accounting Association was reviewing our chartered accountant designation and its place in the modern financial environment. I, along with others holding this designation, would be grandfathered to any future organizational changes. July 1, 2015, I received my new profession certification as a Chartered Professional Accountant (CPA) of Alberta and Canada.

This is to certify to all whom it may concern that

Douglas Kenneth Brewster

was admitted as a member of the
Chartered Professional Accountants of Alberta
on the 1st day of July, 2015,
and is thereby entitled to use the designation

Chartered Professional Accountant

and is awarded all rights and privileges pertaining to the designation,
in accordance with the *Chartered Professional Accountants Act*.

Given under seal of the Chartered Professional Accountants of Alberta

Board Chair

Registration Committee Chair

You might be wondering about who George and Dennis are. Well, Donna and I are grandparents to Bettyanne's three children, Jessica, Callie, and Luke, who have all completed high school and are out on their own. We have great memories and countless pictures of these darlings as they grew up. My son, Stephen and his wife, Amanda, blessed us with two boys, Colton (born in 2011) and Camden (born in 2012). We now have five wonderful grandchildren!

We spend most of our summers at the farm with the two young minions, and it has been great spending valuable time with them, enjoying and spoiling them as we did with the three older grandchildren. Donna loves teaching them about the workings of the farm, and the three of them ride around in the golf cart we bought because of my mobility problems. It's not unusual to see them out and about with the gator, picking up sticks, stones, and garbage in the yard and ditch.

Donna and I take the boys on outings to explore the country through fields, bushes, and water. We count birds and look for moose, deer, and even bald eagles. The seasonal migration of Canada geese and other birds pass over our area where the animals stop for food water and rest.

The return of the children to their mother often involves an apology and a promise not to get their shoes and clothes mucked up again. But the kids are happy about the small rocks I manage to place in their pockets for the drive home.

Now, I have a habit of calling my grandsons different names. So, Colton is also called George. Colton is the oldest. He is quieter and loves playing and telling jokes. He is a lot like his great grandfathers, Ken and Alex, in this regard. He likes order;

he lines up his toys and tells his dad he shut the garage door or put the tractor away. I think, at least at this age, he is a George.

Camden is a Dennis. I think he was put on this earth to bug and disrupt my life. I can so relate to Mr. Wilson. If Dennis is around, I must be alert and prepared for the TV to break down, my alarm clock to go off at 3:00 a.m., or to trip on a toy underfoot.

During a family visit to Calgary one weekend a couple of years ago, Camden fiddled with my recliner, causing me getting stuck. Next, he reprogrammed my sleep machine, and I had to call a technician to get it working again. Then, as he eyed me up and down, I accidentally brushed against a wall picture that immediately crashed to the floor and scared the heck out of me. It took a couple of days after the family left for Saskatchewan for me to settle down and stop worrying about my next accident. So, yes, Camden is a Dennis.

Colton and Camden are always happy to hear my fishing stories, and of course, sometimes I like to exaggerate a little bit. "You know, boys, when I arrived in Melfort, Saskatchewan back in the 1970s for yet another fishing trip, Jerry had every single detail looked after with his fourteen foot aluminum boat ready for me to help flip over and secure on the top of his car. He was a teacher and highly organized," I added.

"Makes sense," said Colton with a nod.

"Yep," agreed Camden.

"Jerry taught northeast of Saskatoon and spent his leisure time touring and fishing the beautiful northern Saskatchewan lakes. He was a real nature lover."

"Did you guys catch anything?" asked Colton, sketching a walleye on the paper Donna left for him and his brother.

"I never caught a giant fish with Jerry. He always seemed to jinx me, especially when trying to haul it into the boat!" I glance down at Colton's upturned freckled face, giving him a slow wink and a smile.

"Right, Poppa," he said.

"Right," echoed Camden.

"You know," I continued, "one time we set up camp on a small island close to Pelican Narrows, known by the Cree as *Opawikoscikcan,* which means "the narrows of fear." Jerry and I ate a good meal that he cooked up on the open fire, and later while fishing on the lake, we chatted about the ingenuity of the Cree in what we now call the Canadian prairies."

"Learning," said, Colton.

"Yes, until…" I said.

"Yes, Poppa? Until what?"

"A bobcat appeared! While enjoying the rich blue colour of the lake water and the deep sunken roots for which the Pelican Narrows is well-known, literally out of the blue a real, live bobcat thrust his sturdy bulk against our boat. His powerful front paws pressed the boat rim far down to hoist himself inside—with us!

"Did you shout for help, Poppa?"

"I said nothing, Colton. My jaws clamped tight. I let my paddles do the talking for me. I smacked the cat with the paddle and then paddled away from its sharp claws with all my strength. Jerry shouted like crazy trying to push himself away from our boat. Not me, though. I just paddled like mad with my oar in the deep water of what the Cree correctly call the narrows of fear.

"Bobcats can float on their backs when they're tired, Poppa. He was just lost and scared and…" Colton trailed off.

"Looking for his home!" Camden finished his thought.

Touched at my grandsons' empathy, I was reminded of Bettyanne and Cathy, two girls with a deep love for animals. Even my Dad was awed by his granddaughters' heart-felt passion for livestock, birds, and pets. No living creature was unimportant to my daughters, a fact crystal clear during Cathy and Bettyanne's many cattle show successes and again when Bettyanne graduated as a veterinary assistant.

"What happened next?" asked Colton, paddling my story back to the boat.

"Well," I said, reliving the memory, "a bad rainstorm hit our little island. Our camp and supplies were still dry, thanks to Jerry, who stored them under the rocks near the trees. Branches waved their frantic arms over the lake water, which slammed against the rocks. We were stranded all right, but far away from the bobcat. We were marooned. The first day was tolerable, the next was downright scary. Jerry and I talked about being at the mercy of the elements. We spoke about the First Nations brilliant forest resourcefulness and inventiveness with fresh respect.

"'Day three, Doug. The savage wind is blasting lake foam,' said Jerry. 'Imagine a life as desperate as this for our First Peoples. I see their courage.'

"Hours later, heavy rain pelting onto the tent and we spied a canoe bouncing across the thrashing waves. In the dark we could see the outlines of a man and a boy. Spotting us, they frantically pulled onto shore. Frightened, cold, and soaking wet, the boy's face was deathly blue. His teeth chattered. Settling them close to a fire he soon had roaring, Jerry piled on the food that they devoured like bears until I noticed with some alarm that little remained. Two hours later, warm and full, the father and young son head home on the open lake. With little grub left, Jerry and I had more anxiety. *What about my Calgary appointments tomorrow?* I thought. *How unreliable does it look not to show up for scheduled business appointments?* I had been groomed to non-negotiable farm-time punctuality, as had Jerry.

"He suggested a plan. 'We plow our boat straight into the waves, so it remains balanced. It will not tip. Then we paddle the shoreline, out of the wind, and toward our vehicle. I'll edge my way around the island in the boat, pick you up, load the craft, and go straight into the wind to cross the lake.'

"His plan worked perfectly until…"

"A bobcat again?" Colton asked, his eyes wide.

"No, the motor quit! We discovered the propeller's shear bolt was cracked. Maybe it hit a rock. I saw the hairline fracture first and looked up at Jerry's bewildered face.

"You know," I interrupted my story, "both people and machines are delicate, eh Colton? Steel is no less fragile than personal feelings, that's for sure." Then I continued. "Blown back to shore once again, we were just about ready to hop onto the rocks when out of the blue…"

"Bobcat?" my grandsons said in unison.

"No, the motor suddenly leapt to life! Our boat bolted into the waves, and maybe four hours later, we staggered to shore in much the same condition as our two guests had the day before. Frozen stiff and starving, we saw where the propeller's broken shear pin shifted, mercifully, to the precise half millimeter needed to propel us across the lake. A minuscule measure of good fortune! Shivering, we dried our clothes and knelt before what I now understood was, without question, sacred fire. While we hunched over the heat in gratitude, I considered my reliable car parked not far away, wheels ready to carry me wherever I wanted to go. I considered the First Nations history and the tremendous innovation of those without the transportation I so enjoyed. Suddenly, Jerry stopped and pointed at an impression in the ash by the fire. We both bent closer."

"What?" asked Colton.

"Bobcat imprints," I said. "Jerry then told me that bobcat prints indicated the balance between body, mind, and spirit. 'Tempered with that steadiness,' he said, 'the lynx signals leadership.' He also told me that when we cross paths with the bobcat, as we did today, it's an invitation to contemplate our capacity for leadership."

"'But, Jerry, the creature we saw was drowning, in trouble, looking for a home,' I said

"'Yes, Doug, a real home, where he could be himself—a place he'd belong and maybe even be celebrated, not a place where he would thrash about in grueling indecision.'

"We examined the paw print in silence until Jerry asked me, 'Ready to move along?'

Looking away from the two boys, I knew back then that my gentle friend was not talking about revving up the car we would soon be driving. I had a personal decision to make about my future and career. That evening, as we drove out of the northern village of Pelican Narrows, in the boreal forest of central Saskatchewan, I sensed I had received a gift. I knew what I had to do next. I knew I would do the right thing for me.

"You know boys, I like to think the bobcat came to our boat with a purpose. That trip caused me to search events for their deeper meanings."

—

Along with the joys of watching our five grandchildren grow and mature and seeing most of our family and friends doing well, there has also been sadness and grief. Donna's sister Susan and her husband, Jim, passed away in 2016 after extended bouts with cancer. It was devastating for everyone, especially for their young sons who were in their early twenties at the time.

In 2017, my sister Donna, passed away at home unexpectedly. It was a shock and reality check for me, since Donna was both younger and healthier than I was. I did not realize how her absence would affect me. Through thick and thin, I depended on her great support and intuition. She was the go-to person for what was happening in Earl Grey and the surrounding areas.

My wife, Donna, had lost her two best friends, each having supported the other through the many crises their families had endured. It was a huge loss, and it took a long time for her to accept that their visits and frequent phone calls were no more, only memories. We think about them often.

Donna began having hip problems in 2015, but with the backlog in the medical system, she waited two years for surgery. The hip replacement, however, resulted in nerve damage in her lower back, which required many hours of rehab and continuing large doses of painkillers. Despite the after effects she experienced from the medication and the numbness in her leg as a result of the operation, she continues her outdoor activities that she enjoys so much, and we continue to be thankful.

—

Relaxing in the yard one afternoon, I heard my grandson call me.

"Poppa, my teacher asked us to talk with our families about the old and the new. We have to compare," said Colton. His vocabulary was impressive for his age. "For something old, I want to study Great Grandpa Ken's tractor," he said before racing ahead. "And for something new and amazing and modern," he continued, his tone rising with excitement, "I want to study my dad's gigantic new green combine. Can you help me on the project, Poppa? Maybe now?"

"For sure," I replied, smiling to myself about which pieces of farm machinery had already won Colton's contest. The two of us headed over to the shed to see Dad's superbly maintained John Deere model 70 diesel tractor, neatly stored in its time-honoured location. I extracted his dog-eared, oil-stained operator's manual, still tucked inside one of the solid wooden dog cages he built for me back in 1952. Those crates now serve as sturdy tool cupboards.

"Here, Colton. Here is the 125-page service manual from 1955. At the back is an index to look up every part of your great grandad's diesel-run machine. Look," I said, flipping the pages, "you can see my dad's oily thumbprints on some margins more than others. What does that tell you?" I asked my grandson, his animated face still riveted in the direction of the glassed-in cab of his dad's state-of-the-art new and computerized green John Deere #770 combine.

Glancing through the smudged instruction guide in my hand, it is easy to trace the upkeep history of a tractor my dad was proud to own.

"My dad worked hard all his life, Colton," I said, noticing the overlapping prints that testified to heavy use of a page entitled 'Lubrication Chart for General Purpose Tractors.' The instructions advised a mixture of oil for brakes, along with a warning that lubrication with the wrong weight of oil could result in loss of power, excessive fuel consumption, undue wear on moving parts, and the eventual replacement of costly parts.

"The fear of breakdowns," I told Colton, "was almost as scary as fire, especially for folks so far away from a repair place or a fire station. Remember, there were no computers at that time—no Googling instructions or You Tubing video classes available in your great granddad's time. In those days, there was only a battery-operated wall phone, and the phone line was shared with the neighbours. That meant that if one of the neighbours was busy talking on the phone, we had to wait until they were finished before the line was usable."

"No iPad?" asked Colton

"No. Isolated farmers like my dad had to know their machines. They became experts in ways to care for them, now known as preventable maintenance. They had to educate themselves. There was a vast wealth of knowledge shared between farmers back in the 1950s and 1960s, especially those living far from any service outlet."

"Here's a little story, Colton. One fall day after harvest, a sudden rainstorm soaked the topmost bales we'd piled high in the field all morning. I saw Dad's sunburned forehead wrinkle up with thick lines of distress. The heavy rain had spoiled the top bales. Then, the drenched wheat bales began to expand, breaking the tie strings with the weight of the rainwater. Disheartened and knowing the mess of bale and straw would have to be cleaned up later, we headed home."

"Would Grandpa have balked at the expense of automatic plastic covered round bales, Dad?" Stephen asked, joining our impromptu gathering.

"Yes, I think so, son." I said. "He'd already built the storage structures he needed for his grain production, so he might have calculated the extra expense would eat up his profits. For farmers today, though, the automatic bale-wrapping of dry hay minimizes the effects of bad weather. Wrapping the hay out in the fields eliminated building more expensive storage."

"Yes," said Stephen, shaking his head. "Plastic bale wrapping means cattle farmers and ranchers no longer worry about weather conditions like Grandpa did. Back in the day, the words 'wet hay' foretold disaster. No one buys or even feeds spoilage to their cattle."

"But Grandpa, plastic is made of oil!" said Colton. "Ms. Wright says we must stop polluting the land and oceans with it. It's choking the fish!" He then told us about the video of a fisherman gently extracting a plastic straw from an innocent turtle's nose. "Once it was out, the poor turtle blinked and breathed again," said Colton, still gazing out at his dad's mighty green machine.

"Colton," I replied. "Just as tractors improved farming when they replaced horse-drawn machinery, bale wraps improve life for modern farm owners. But I hear you, Colton. I hear and respect your concern, and you are right. Even as we improve, we must also take responsibility for the planet."

"And the turtles, Poppa," said Colton, a proud smile on his face.

"Look," I said, giving him a graphic example instead of using big words like 'incalculable damage.' "Remember last night when you wanted to show Nana the wonderful Lego castle you'd built?"

A sudden frown formed on Colton's smooth face. "Yes, Mady crashed right through it! He ran to Nana right through all my work!" Colton's lower lip began to tremble at the memory. "It's wrecked!

"So, what would a solution be?" I asked. "Should you not allow your dog in the house anymore, or is there another idea, Colton, about ways to protect your hard castle-building work?"

"Well, I guess I could build it on a table?" he said.

"Sure! And your great grandpa Ken felt just as upset about the ruin of his bales, too. But his only option was piling the bales out in the field, and it wasn't working for him. The way forward, Colton, is to think about new solutions, like you just did by deciding to work on the table instead of the floor. When a new opportunity presents itself, welcome it!"

"Dad says you keep your eyes wide open for new ideas, Poppa," said Colton. "Me too, Poppa."

What a pleasure it was for me to feel such closeness with my grandsons.

"There are now smart, innovative companies, Colton, that recycle bale plastic into home building material. Decks and cabins and solid fences are being made of used bale plastic."

"Oh, great!" said Colton.

"Maybe you can tell Ms. Wright that no farmer wanted to bury or burn used bale plastic or grain bags, but there was no place for disposal at that time. Now, they are rolled up tight and shipped for recycling. So, farmers get to have their hay and grain covered without affecting the environment."

"This waiting business," said Stephen, looking first at the grey sky and rain clouds and then at his phone, "is not easy." He repeated it as he jogged inside to get us each a glass of water.

"Our new up to date technically operated combine has stopped, Dad," he said, his expression grim. "Now I'm waiting for a service pro who should arrive," he glanced at the time, "within the next three to four hours. Texting him was easy, but his getting here in person is the same old slow," said Stephen.

"Want to hear an old story of another stall?" I said with a smile, pulling up another old dog crate seat. "Looks like we have time on our hands."

I smile as Camden and Colton climb onto their father's lap, and I begin my tale.

"I was a young boy, like you two characters, being already too big to crawl between stuck blades of stalled machinery. One humid afternoon, I rode with my dad on the tractor and we got stuck in the mud. The wheels turned, but we did not move. We got off the tractor, and Dad investigated the problem. He did not want to bother our neighbour just over the hill.

"Without asking Dad's permission, I jumped up and ran. In less than five minutes flat, Mr. Brandt grabbed his home-made clippers, welded and wired to extended wooden poles, and hauled me onto his horse. We charged back to Dad. In a single fluid motion, Mr. Brandt slid from his horse, sunk the scissors deep into the mud under the spot Dad indicated and began chopping at what could not be seen. Long minutes later, I heard, or maybe felt, the smallest release and the tractor's tiny clunk forward. Seconds later, we viewed the ancient tree roots that lassoed our modern machine to a standstill.

"'Start it, Ken,' Mr. Brandt said, and low and behold, the tractor hummed and moved ahead as if nothing had happened.

"Before Dad could jump down to thank him, our neighbour was galloping back to his own tasks. His homemade appliance, clumped with mud, bounced against his saddle belt.

"'Wonderful neighbour,' Dad said. His eyes sparkled moist before he quietly climbed back up on the tractor with me."

My grandsons had listened intensely, their eyes sparkling. Stephen glanced over at me, without the boys seeing him, and to motion me as if to say, "Now that sounds a little farfetched."

"Dad!" Camden urged his father. "I can do what Poppa did. I want to run over and ask the Kenan's to come to help you! I know they will!"

"Yes, they sure would, Camden, but the fact is, they wouldn't know what to do. They are not trained specialists. The solution requires a schooled computer modem technician specialist from John Deere," said Stephen.

Stephen searched "702962," the serial number of the old John Deere 70 diesel tractor, in the search bar of the John Deere website. I considered that my son, his wife, and their children are continuing, improving, and expanding the farm operation that began with four generations of farming in Saskatchewan with my grandparents, Hazel and Maurice Brewster, moving from Montana to Earl Grey, their Canadian homestead. Our prairie history is so young, especially in our area, which was pioneered in the 1880s, not even twice my lifetime ago.

"Nothing for Grandpa's tractor, Dad? Is it too old?" Colton asked.

"Boys," I offered. "How about while we wait for the service specialist, we take my dad's old tractor out for a bit of a run?" With a nod from Stephen and a yelp from my grandsons, I settle onto the padded steel seat. Ready. The little gas motor revved up noisily, and cranking the diesel motor over, it started. Then it spat the distinct sound of the two-cylinder diesel: *putt putt putt*. Once one hears a two-cylinder designed engine, one never forgets it.

Back in the day, I had my wide-brimmed hemp cloth hat to prevent third-degree sunburns to my ears every spring season and slipped a jar of water into my lunch bag. In those days, if not hand pumped directly from our eighty-foot-deep well, we drank warm water that got even warmer as the day went on. I am now told that drinking warm water is much healthier.

Cautiously, I steered Dad's tractor down the road into the adjoining field. Now, even my youngest grandchild had had a turn at driving the historical John Deere 70 diesel tractor. The ride reminded me of a speech I once made to young farmers where I celebrated the fact that power steering was the single-most magnificent invention that modernized farm equipment. Extremely enthused, I shared that Dad's self-propelled John Deere #65 combine, purchased in 1961, made his arms so sore that he had to sell it. Brutal fatigue brought on by turning and hitting field rocks and ruts strained Dad's arms and back, which caused physical pain later in his life. But no more! After power steering took hold, dedicated farmers like my dad were able to extend work lives, which is what they wanted. They wanted to farm year after year, and power steering meant they could hold their place on the land.

Today, computer panels, rotational sign lights, covered equipment cabs, lifesaving air-conditioners, chill-busting heaters, and state-of-the-art music sound systems make a farmer's work much easier and safer. However, all of us, then and now, were and are dedicated suppliers of food in Canada and around the world.

PRESENT TIME

TWO THOUSAND AND EIGHTEEN WAS A GOOD year for me. At Dr. Isaac's forceful suggestion, I lost twenty-five pounds, and I felt so much better. My sugar levels stabilized, and I was able to walk more and do things I have not done for years. The routine of watching what one eats, including smaller portions, and walking a kilometre a day paid in dividends.

I continued to enjoy operating our D6N dozer, clearing bush on land near the Bundus farm—the farm Donna and I travelled to one winter night forty years ago. As we were there this summer day many years later, a nice yellow lab appeared, looking up at me as if to ask how I was. Content, he then ran around to look for whatever he could find. I was concerned about hurting him with my big machine, but then I remembered, he is a farm dog. I stopped to pet him; however, he was too interested in enjoying the daily activity and surroundings. He scampered off, but I was sure he would be back later after checking out his home turf.

Upon leaving the field that evening in my old service truck, I passed through another neighbour's yard and was greeted by another friendly face. This time a black lab approached, wagging his tail. I was familiar with this dog; he would visit our grain bins next to his yard. I stopped and opened the door for him to come up for a little attention before soon going our separate ways.

Heading home to a late supper, I could not stop marvelling at these typical farm dogs. Our family has had some wonderful farm dogs over the years; my favourite pet was a big black border collie named Bullet. He grew up with me and was there for me when I was young. He loved riding on the open tractor with Dad for hours or running behind the implement. A farm dog just needs a nod or glance to know he is wanted and loved, for his confidence to be reassured.

—

The campground and the Saskatchewan farmyard continue to be improved. Along with Earl Grey's farm shop, Douglas Industries Ltd.'s old manufacturing building underwent a major renovation.

With two recent family weddings, our children and grandchildren are living their lives, and we enjoy their visits. We are content.

In November, I received a letter from the Chartered Professional Accountants of Alberta, congratulating me on my fifty-year milestone as a member of the Association. It is hard to comprehend that it has been fifty years since I wrote my exams at the University of Calgary. I was gifted a bear sculpture, which I appreciated. It is said to represent the strength, self-reliance, and expertise I have displayed over the last fifty years. What a thoughtful gift.

November 26, 2018

Mr. Douglas BREWSTER, CPA, CA
35 Tuscany Hills Park NW
Calgary AB, T3L 2A2

Dear Mr. Brewster:

The exceptional reputation of the CPA profession in Alberta is a product of remarkable professionals like you. From contributing to the workplace to volunteering for the profession, from making a difference in the community to helping build the economy, Alberta's CPAs like yourself are making a difference.

On behalf of CPA Alberta, we congratulate you on your 50-year milestone as a member of the profession.

As a token of our appreciation for and acknowledgement of your accomplishments and contributions, please accept this gift to commemorate your milestone. This distinctive cast aluminum sculpture is called "The Bear," and represents the strength, self-reliance and expertise you have displayed throughout your 50 years in the profession. I hope you can find a prominent place to display this memento!

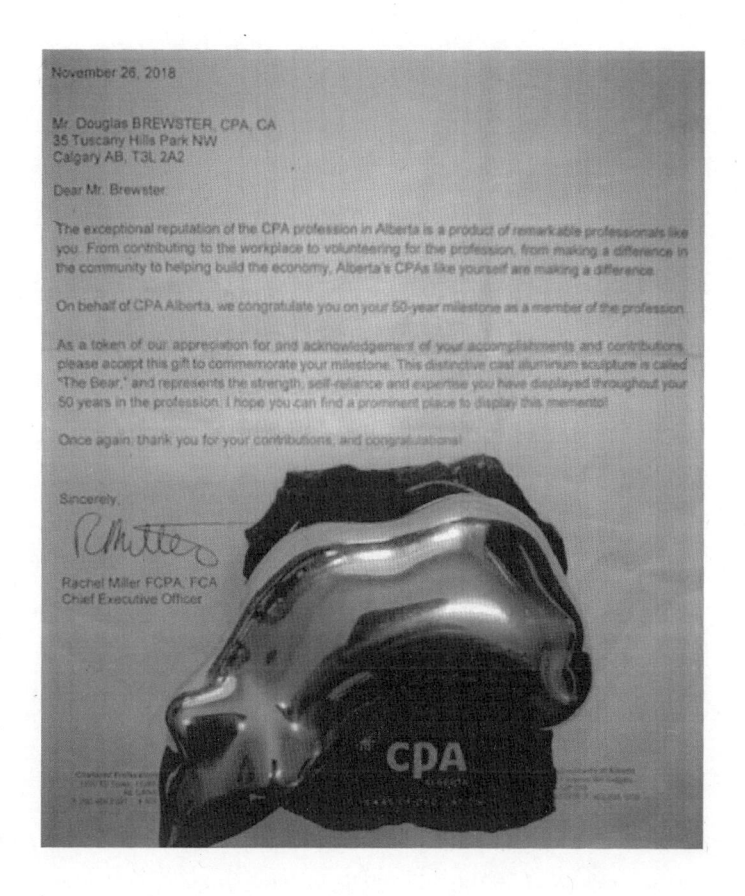

Once again, thank you for your contributions, and congratulations!

Sincerely,

Rachel Miller FCPA, FCA
Chief Executive Officer

—

In 2019, we enjoyed our return trip to Palm Springs and Oceanside, California for a winter break. I continued my daily walks and had an excellent report from the transplant team on my annual spring review. Donna could not wait to return to volunteering doing Income Tax at the Kerby Centre.

With spring's arrival, we returned to the farm and our little house in Earl Grey, Saskatchewan. The boys, Colton and Camden, are growing up and greatly help their mom with her busy days. Bettyanne and her family still live in Yorkton and are doing well. We are fortunate to be able to spend a week camping at Madge Lake with them. Even more, we became great grandparents with the arrival of Ivy on June 21, 2019. Proud parents Jessica and Danny were thrilled. But this old guy couldn't help but wonder, *do they have a clue what lay ahead of them?* Luke is busy enjoying freedom from school and having a job that provided him with his own money to spend. Callie has a summer job and is getting ready for her fourth year at the University in Regina.

After a terribly dry spring and early summer, the moisture came and never quite left. From August on, the weather was consistently cool and humid. Harvest was late and caused huge problems for the farm community. Unlike other years, grain did not ripen or dry down, and with snow coming early, combining had to wait until next spring, which resulted in substantial economic loss for many. Our farm was fortunate to have a grain dryer, and Stephen has spent many hours operating it, with me running his combine more than usual.

Donna and I were fortunate to get our motor home ready for travelling, not just sitting at our campground. It had been our home for the better part of 2005 to 2007, along with being our summer home in Saskatchewan until 2010. We are so comfortable in it!

Donna and I headed back to Calgary for the winter and eagerly awaited our family making the long trip to join us for Christmas. The Christmas season is a special time for Donna and I, and it will be great being all together this season. We are in good health, and we are so thankful for what we have.

The year 2020 arrived with many new and unexpected happenings. Our family had a great Christmas holiday in Calgary, and everyone returned home safely. However, my aunt Jean Brewster—my Grade 1 teacher, my favourite aunt, and the last one of her generation—passed away in early January, so we returned to Saskatchewan for a few days. We had planned a two-week winter holiday to warmer weather, which we then changed. Reality and our past were at the forefront of our new year, 2020. The optimism and confidence we had built up over the last couple of years was quickly being eroded, starting with the emotional effect of Aunt Jean's passing, the stressful travel schedule, and my becoming ill with fluid in my lungs while sitting in Florida. Donna was able to get us an immediate five-hour flight home, where I received medical help. We were so thankful to be back on Canadian soil after such a long and stressful trip, and thankfully I did not need to stay in the hospital this time.

Within days, the world started experiencing a global pandemic (Covid 19), jolting us all. My reality of a compromised immune system and having had a heart transplant will require me to be extra careful. It was another reminder of how lucky I am and have been. I wake up each morning, and time goes on.

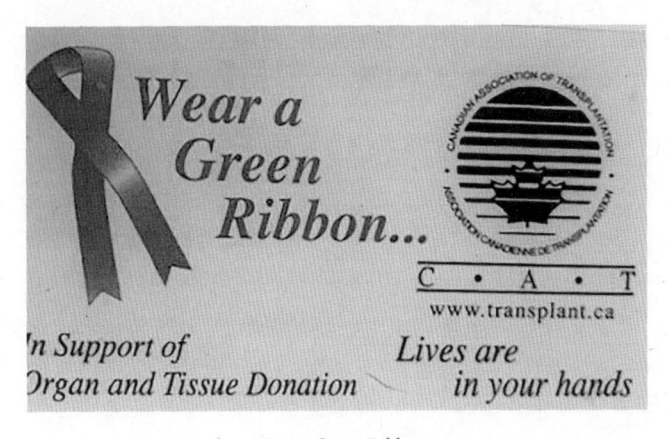

Wear a Green Ribbon

ACKNOWLEDGEMENTS

TO DONNA, MY WIFE AND PARTNER, WHO has not only given up her precious personal time over the past many years dealing and helping with my health issues, but helping record and put my life story to text. Donna's ability to document, organize and edit made this memoir a reality of which I am so indebted and not able to thank her enough.

To Eleanor Cowan, BA, B.Ed., writer, and author, who helped greatly with bringing the story to life through many ideas and the initial writing and editing.

To everyone who inspired me to tell my story in the context of a period in history that was my time. I am indebted to so many, and I ask that you take satisfaction in knowing we did this together.

To those who helped me gather materials and information from far and wide. I thank all who provided, directly or indirectly, information, photos, and poems.

To the communities of Earl Grey and Gibbs, Saskatchewan who provided me with an environment specific to rural Saskatchewan during early times in the province's history.

To the short stay in the winter of 1948, along with summer visits with Mom (in my early years), to Peterborough, Ontario—Mom's birthplace. Although short, it left strong impressions of kindness and of a much simpler kind of existence.

To the province of Saskatchewan, I owe a great deal of gratitude for providing an atmosphere for education and work, which played a pivotal part in my life.

To the province of Alberta, which has provided me with many opportunities, including meeting my wife, Donna. I have experienced the better part of fifty years of the province's highs and lows of its economy. Alberta allowed me opportunities that I had thought were simply impossible.

Numerous sources were used to help understand history on the prairies in preparing my memoir, including, but not limited to:

From Buffalo Grass to Wheat
by Leonard Sheils

Harvest of Memories by Earl Grey History Committee

Silton Seasons by R.D. Simon

The Brewster Genealogy: Volume 1 and Volume 2
(1566-1907) by Emma C. Brewster-Jones

Printed in Canada